Choosing Solar El

A g

PET
2020 ∝

Bria

20/4/19

621.31244

© Centre for Alternative Technology, 2010
by Brian Goss;

Centre for Alternative Technology,
Machynlleth, Powys,
SY20 9AZ, UK
Tel. 01654 705980 • Fax. 01654 702782
www.cat.org.uk

ISBN: 978-1-902175-59-1

1 2 3 4 5 6 7 8 9 10

Publisher and lead editor: Allan Shepherd
Editors: Caroline Oakley, Helen Theaker, Hele Oakley, Lesley Rice
Typesetting and design: Graham Preston
Cover design: Annika Lundqvist
Symbol design: Annika Lundqvist
Cover graphic ©: Savio Alphonso

Additional text about feed-in tariff incentives appears courtesy of Tobi Kellner.

All images and symbols belong to © Brian Goss or © CAT except where credited.

Other images appear courtesy of © Solar Century, © Dulas Ltd., © Rob Gwillim
© Dr Ralph Gottschalg, © Max Fordham & Partners, © Solar Integrated
Technologies, © Effergy, © Norsksolkraft and the © IEC

Published by CAT Publications, CAT Charity Ltd., Registered Charity No. 265239

Printed in the UK by Cambrian Printers, Aberystwyth. 01970 627111

Contents

Preface

Who should read this book?

Who should read this book? In short, anyone who wants to use solar photovoltaic (PV) to generate electricity and needs to know the pros and cons before talking to a professional solar installer. *Choosing Solar Electricity* is not intended as a comprehensive technical manual for designers or installers, although those considering installation as a business will find it a useful quick introduction to the subject. It doesn't cover design and installation issues exhaustively, but does give you the knowledge to feel confident enough to engage an installer. It also helps you understand how feed-in tariff incentives could make your investment in a solar PV system more than pay for itself. There are many books about solar PV technology and systems in print. Most make the assumption that the reader has a scientific background and a prior understanding of the principles of electricity; that they can easily understand concepts by looking at algebraic equations. My experience of teaching short courses about solar electricity has shown me that people interested in the subject come from a variety of backgrounds. Some may have qualifications in electrical and electronic engineering, whilst others have not studied science since the age of sixteen. And whilst some wish to pursue PV design and installation professionally, or have some involvement in design and installation, many may find that the safety and competency requirements for electrical work make DIY PV installation an unrealistic proposition. Instead they choose to have a system installed professionally. Homeowners, estate managers and building services managers who choose to have a system professionally installed will nonetheless want to understand what they're being offered and to make well informed design decisions that affect the aesthetics and performance of their new PV system.

How to use this book

Because the technical nature of solar PV can be daunting, I've tried to make things as simple as possible by providing easy to read break-out boxes for safety issues and technical boxes for those who want more in-depth technical explanations. You can choose to read the book without dipping into the technical boxes, but reading them will give you a greater depth of understanding. A professional solar PV installer will have a thorough knowledge of safety issues but as an operator of a PV system you will need to be aware of these too.

SAFETY ————————————————

Expert advice should be sought before having a solar electric system installed. Check out Chapter Ten, 'Finding an installer'.

TECHNICAL ————————————————

You'll find boxes like this one throughout the book. They have extra detail for the more technical reader; someone who has a greater interest in the system they're having installed. If you're not so technically inclined, don't worry, it's not necessary to know these technical details to own or run a PV system.

A step by step approach

Working out what you want from solar PV and what solar PV can give you requires a logical step by step approach. Going into it without a thorough understanding of your own requirements and the limitations and potential of solar power could cost a great deal of money. Installing any new power supply is an investment. You'll want to get a good return from the money you spend. Although solar panels have come down in cost and are more efficient than they once were, they are still a pricey option if you get it wrong. Your intended site might not even be suitable for solar, a fact best worked out before you start spending money. The process of getting a PV system installed is outlined in the flowchart below – the various stages will be explained throughout the book.

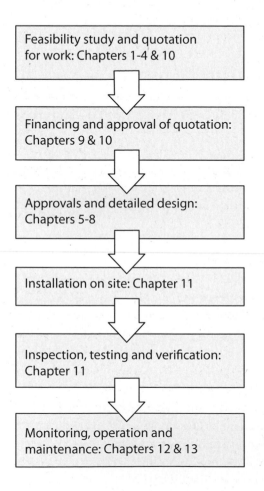

Fig. 1. Production flowchart.

Chapter One

Choosing renewables

Introduction

Although this book is about solar PV it is useful to know a little about the other renewable technologies and how solar PV compares – and can work in unison – with them. Many people chose to combine solar PV with other renewable technologies to provide a complete green solution to their heat and power needs. It is worth keeping this in mind when taking decisions about how much money you want to invest in PV. Choosing technologies carefully will help pay back your investment quicker.

Heat versus power

Generally speaking individual renewable energy technologies provide either heat or power (in the form of electricity). When we want to produce heat we turn to solar water heating, passive solar space heating, ground source heat pumps and wood burning stoves (pellets, wood chip or logs). When we want electricity we pick solar PV, hydro or wind. There are some renewable energy technologies that are designed to provide both heat and power. These are called combined heat and power (CHP) units and are generally used to meet the heat and power requirements of a large number of users – for example, a public or private office building, a university, a housing estate or a village. These systems are beyond the scope of this book.

How do I know which renewable energy system is right for me?

The most important issue when considering a renewable energy system is choosing the right technology for your situation, whether related to your site, the state of any existing technology you wish to replace, or the way you wish to use renewable energy.

The site

Any decision you take will be very site specific. Site requirements for solar PV systems – described in detail in Chapter Two – are very different to those of other renewable technologies. For example, a small wind turbine will produce minimal electricity if installed on a building within a large housing estate because such places usually have calm, irregular and turbulent winds, which is just about the opposite of what a wind turbine actually needs. To recoup installation costs a wind turbine needs strong, regular and consistent wind speeds, which is why wind farms are sited out at sea or on hilltops. Likewise, you shouldn't consider micro-hydro power (that is, domestic scale water power) if you don't live next to the kind of fast flowing streams typically found in hilly areas; or a ground source heat pump if your house sits on granite (unless you like playing with dynamite!).

Existing technology

It's very easy to get excited about buying a renewable energy system, but you should assess whether you really need one. For example, you might not want to convert to a wood pellet heating system if your gas boiler is only two years old. It will be performing very efficiently already. Likewise, if you are already connected to the grid you might be better spending your money on energy conservation measures and buying green electricity from a national energy supplier. On the other hand feed-in tariff incentives – with a guaranteed price for every unit of electricity generated over a 25 year period – make solar PV a realistic investment opportunity in its own right (see Chapter Nine).

Appropriate use

Make sure the technology fits the use. Are you choosing solar because you want to reduce your carbon emissions, get a good return from a long term investment or because you need a backup in the event of power cuts? If it is the latter, note that most European grid-connected PV systems are not designed to operate during power cuts (since power cuts are very rare in most European countries). PV systems can be installed with battery backup, but this adds heavily to their cost and environmental impact (see Chapter Five). Hence for the backup of a utility supply that only fails for a few hours a year – if you really think you need one – a small UPS (uninterruptible power supply) or petrol generator often makes better financial and environmental sense.

Put energy conservation first

Near, if not at the top of your list of considerations should be whether renewable energy is your first priority at all! You will usually get a better return on your investment by installing basic insulation and energy efficiency measures if you haven't already done so. However, if your home or building has energy efficient heating and lighting and is well insulated already, then on-site renewable energy is a logical next step in your drive to reduce your energy consumption and cut your carbon emissions.

Choosing solar

Solar is a versatile energy resource and can be used in one of three different ways, although in this book we are concerned only with the third:

- To create so called 'passive' heat – where glass and careful design elements control the light and heat of the sun as it passes into a building.
- To heat water. Solar thermal collectors usually have water pumped through them, which is then heated by the sun. This hot water is used to heat water in a copper cylinder and supply domestic hot water (DHW).
- To produce electricity. Solar electricity, or photovoltaic (PV), modules convert light into electricity (*photo* = light, *voltaic* = electricity), which is fed directly into a building's electricity supply.

Although passive solar is an important component in architectural and building design, most people considering solar power have the two 'active' solar devices – solar thermal collectors or solar PV modules – in mind. Both are often called solar panels, but it is best to avoid this generic and potentially misleading term as it can cause confusion between water heating and the production of electricity.

Some people ask: 'Why can't we use electricity from solar PV to heat water?', or, 'Why have two separate systems?' Well, the answer is: technically you can have one system for both heating water and producing electricity, but it would be very inefficient and costly to do so. Much better to install a solar thermal collector and save the solar electricity for equipment which can't run off heat, like: power tools, computers, lights, televisions, and so on.

Why solar PV and not wind or hydro?

Solar PV can be used to provide power for homes, businesses, public buildings, holiday cottages, boats, schools, electric vehicles, remote devices... even satellites in orbit. Wind and hydro can do most of these things, too (except the satellites!), so why pick solar?

A major advantage of solar PV is that it is much less site-dependant than wind or hydro. You don't need a stream or an open windy plateau and, unlike a wind turbine, a solar PV array can be efficient when placed on an urban roof. The one issue you do need to be careful of is shading. A PV system will perform efficiently if it is placed with an open vista to the south, but not if it is heavily shaded by a large building or tall tree. Working out whether or not shading will stop your solar PV system being an economic proposition or make it slightly less efficient than it might otherwise be is your first priority. To do this you will need to calculate your solar resource. The next chapter shows you how.

Additional resources

It is beyond the scope of this book to offer a detailed comparison with other renewable energy technologies or show you how to use solar in conjunction with them. For this information, read *Off the Grid*, *Choosing Windpower* and *Going with the Flow* (all published by CAT: www.cat.org.uk). It is quite possible to design a system with all three power sources, but this scenario is more common in remote, off-grid locations. Furthermore, homes already connected to a utility electricity supply are more likely to use solar or wind power due to the remoteness of streams suitable for hydro schemes.

Chapter Two

The solar resource

Introduction

This chapter will show you how to assess your solar resource – in other words, the amount of solar energy that is available at any given site – to convert into electrical power. It is important to assess your solar resource before you do anything else because it really will define whether or not you can create a useful solar PV system. And by useful I mean productive and cost effective.

If your site really is in the wrong place for solar it is best to know now before you get on to costing and system design. If you haven't got a site in mind yet you can follow the exercises in this chapter anyway. In many ways, it is useful to pick out two or three trial sites to get a good understanding of the different factors that might alter the solar resource.

As mentioned in the last chapter, shading is the primary concern, but there are other important things to consider too: the angle of the roof, the latitude of the location, variations in weather conditions. I will go through each of these points as we go along.

Tools

You'll need a few tools if you want to do the exercises in this chapter for real. Obviously there is a cost involved in buying these tools, so you might prefer just to understand the process rather than complete the project. Your PV installer will do this assessment for you anyway.

TOOL BOX

Compass

A compass enables you to determine the direction in which you are facing, measured in degrees. You'll need a good orienteering compass or a sighting compass: available from most good outdoors shops.

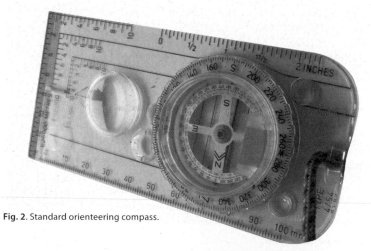

Fig. 2. Standard orienteering compass.

Fig. 3. Sighting compass.

Inclinometer and spirit level

An inclinometer allows you to measure the pitch of your roof. It can also be used to measure the elevation angle of objects that might cause shading of a PV array. You can also measure the pitch of your roof with a spirit level if it has an angle finder.

Fig. 4. Electronic inclinometer.

Fig. 5. Spirit level.

Hand-held solar horizon estimator

These are more intuitive to use than the compass and inclinometer options. However, they are expensive to buy and don't have any other uses, unlike the compass which can also be used to measure the azimuth angle of the roof, and so on.

Fig. 6. Using a solar horizon estimator.

A quick guide to understanding electricity

Before we go any further it will be useful to provide a quick guide to some of the terminology you will come across in this chapter and the rest of the book.

Watts and power

All solar PV modules have a power rating measured in watts (W). If a module is rated at 100W it means it will produce 100 watts of electricity during (and only during) optimal conditions. Power is the rate of delivery of energy so the power rating of a 100W solar PV module is telling us that it will theoretically give us enough electricity to run nine low energy 11W bulbs at the same time, but only when the sun offers peak irradiance (which I talk about below). This is only theoretical because all electrical systems lose some power as the electricity travels through cables and other system components.

   Straight away you can see the value of energy conservation. Our solar PV module will produce enough energy to allow us to run nine low energy bulbs compared to just one and a half conventional 60W bulbs. As solar PV modules are expensive, the cost savings of energy conservation become very apparent very quickly. As a result, owners of solar PV systems tend to become more conscientious about how they use electricity compared to the time before they had a solar installation.

Solar PV array

When several PV modules are installed together we describe them as an array. The total power of the array is simply the number of modules multiplied by the power of each module. So an array consisting of ten 150 watt modules would be a 1500W or 1.5kW array.

Working out the solar irradiance

The intensity of solar radiation falling on a surface is called irradiance (abbreviated to the symbol G). As you'd expect, irradiance is the variable that most affects the output of a PV system – though there are many factors affecting irradiance itself.

• Latitude and air mass (the optical path length for light as it travels through the atmosphere – see technical box on page 14).
• Weather conditions, that is, cloud cover.
• Shading.

The irradiance (or intensity of solar radiation) at noon during good weather conditions is approximately $1000W/m^2$ (watts per square metre). This figure is important because it is the benchmark used to rate the output of any given PV module. For example, if you buy a 100W solar panel, it will only produce 100W at $1000W/m^2$, and hence less than 100W if there is more cloud cover or the sun is lower in the sky. Note that in North America irradiance is often called insolation, so the term insolation is often used on websites.

Energy and power

We use the term 'watt' to describe solar irradiance per square metre, not 'watt hours'. This indicates that irradiance represents power *not* energy: be careful not to confuse these two quantities.

Energy is a finite amount of ability to do work (either stored up in the sun, or in batteries). For example, if you eat a chocolate bar containing 500 calories, your body might burn those 500 calories of energy during a 40-minute bicycle ride. The calories have given you the energy to do the cycle ride.

Electrical energy is measured in kilowatt hours (kWh). Power is the rate at which energy is used or converted. For example, a 1kW electric heater will convert 1kWh of electrical energy into heat energy every hour. A kWh is the same unit of measurement recorded on a household electricity meter or electricity bill, where it is often just called a 'unit' of electricity.

Therefore, if 1kWh of solar energy (or radiation) falls on one square metre ($1m^2$) of solar module in a given time – for example a day – then we would say the solar irradiation is $1kWh/m^2$. Hence, solar irradiation is measured using the unit kWh/m^2 (kilowatt hours per square metre).

Factors affecting PV system performance

This chapter will show you how to get the most value from solar irradiance so that you can convert as much of the sun's energy as possible into usable power. It's worth reiterating that the performance of a solar PV system will be affected by:

• Latitude and air mass
• Cloud cover
• Shading
• Tilt
• Orientation

There is not much any of us can do about the first two, and the other parameters can only be influenced for new buildings or major renovations. Sadly, there is no requirement in building or planning regulations to design roofs with optimum tilt and orientation for future solar installations.

In new build it is possible to maximise the efficiency of solar irradiance by correctly orienting houses, keeping shading to a minimum and building roofs or alternative solar PV sites with optimum tilt and orientation. This does not rule out solar for those houses with less than perfect conditions, but it may make it less efficient and consequently more expensive (in terms of the cost of each unit of electricity produced).

Latitude and air mass

The irradiance reaching a site is affected by air mass, which is determined by the latitude of the site and the time of year, as well as cloud cover and shading (see box below).

TECHNICAL ————————————————————

Air mass

Air mass (AM) is the amount of dust and gas particles the sunlight (photons) has to get past on its passage through the atmosphere. Air mass increases when the sun is lower in the sky. In other words, it increases towards dawn/dusk and is more pronounced the further you are from the equator (that is, it increases with latitude).

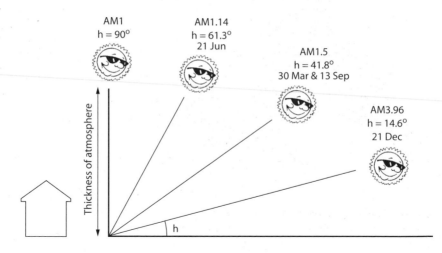

Fig. 7. Showing the effect of solar elevation on air mass at different times of year.
Image courtesy of Dr Ralph Gottschalg.

On a sunny day with clear skies sunlight takes a relatively straight course through the atmosphere, called direct radiation. If, however, the sky is overcast the light is scattered by the clouds, called diffuse radiation. Also, some sunlight will be absorbed by clouds or reflected back out of the atmosphere, and so in general, solar radiation is less when there is thick cloud cover. But, just as you can still find your way when it's cloudy, a solar module will still produce solar power, just less of it.

Also note that because the irises in our eyes compensate for variations in brightness, the difference between bright and cloudy days is actually greater than we perceive it to be. So, if your property is located in an area of mountain weather, with frequent cloud cover, the annual solar radiation will be less than that at the coast where clear skies are more prevalent, as shown in figure 8. A solar module will also produce more power when the sun is high in the sky, that is, in summer and in the middle of the day.

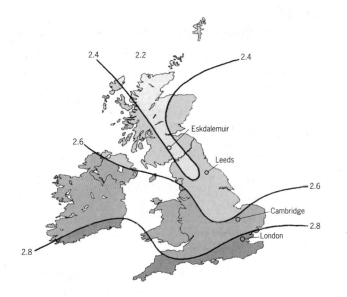

Fig. 8. Average annual solar radiation (kWh/m²/day) for Great Britain and Ireland. *Image courtesy of Max Fordham & Partners (1999) 'Photovoltaics in Buildings: A Design Guide' DTI URN: 99/1274.*

Shading

Incoming irradiance can be restricted by shading from hills, trees, buildings and even roof features such as chimneys and dormer windows. Shading of even a small part of a solar PV array can affect the output of the whole system because of the way the modules are wired up, so careful design is required.

Shading objects of a given height will evidently have more of an impact on an array when the sun is lower in the sky, that is, to the east and west (in the morning and evening) and during the winter. The complex variations of the sun's path can be clearly seen using a 'sun-path' diagram.

Sun-path diagrams

As the name suggests, sun-path diagrams are a handy tool for working out the 'trajectory' of the sun in relation to your proposed solar PV site. The sun-path diagram helps build up a picture of how much irradiance will be restricted by obstacles at different times of the day and on different days of the year. As it has been said, although a solar PV module may be rated at 100W it will only achieve this output under ideal conditions. At all other times it will perform slightly less well. How less well will determine the costs of your system in relation to the power you get from it.

In a typical sun-path diagram, the *y* (vertical) axis represents the elevation angle of the sun. The *x* (horizontal) axis represents

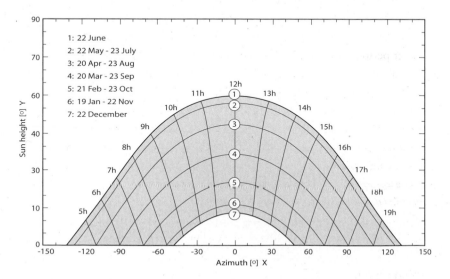

Fig. 9. Sun-path diagram for a location at 53.6° latitude 3.8° longitude.
Image courtesy of Max Fordham and Partners (1999) 'Photovoltaics in Buildings: A Design Guide'
DTI URN: 99/1274.

the azimuth angle – the horizontal angle between due south and the position of the sun at any given time.

It is worth noting here that the azimuth angle is taken with 0° as south whereas the bearing taken on a compass is taken with 0° as north. So you need to add 180° onto the azimuth angle to convert to a compass bearing and subtract 180° from a compass bearing to get the azimuth angle as shown on the azimuth scale on the sun-path diagram opposite.

The movement of the sun at different times of day and during different seasons The left-right curved lines on the figure opposite represent the arced course of the sun as it passes through the sky at different times of the year. Look at the key and you will see line 1 represents the movement of the sun across the sky on June 22nd (the longest day), Line 2 the movement of the sun between 22nd May and 23rd July, and so on right down to line 7, which is the movement of the sun on the shortest day – December 22nd. The further into winter you are the lower the sun is in the sky and therefore the less solar irradiance there is. Therefore, it is obvious that in winter you will get much less power from a PV system, especially if there are lots of shade obstacles in front of your array. However, for most grid-connected systems, because the winter period only accounts for a small proportion of the PV system's annual output, a performance reduction during this period has much less of an impact on the overall performance than a reduction during the summer. In contrast, for a stand-alone system the winter performance is a high priority in high latitudes, since this is when electricity demand is greatest due to the long dark evenings.

The up-down radial lines on the sun-path diagram represent times of day Again, you can see that shading is most likely to be a problem at the start and end of the day, while the actual amount of energy available is smallest during the shortest day (line 7) and greatest during the longest day (line 1), as you would expect.

Measuring the effect of shading using a sun-path diagram
The effect of shading can be measured by plotting your array location on a sun-path diagram using a compass and inclinometer. To do this you'll also need a sun-path diagram print out, a pencil and an eraser (everyone makes mistakes!). Before you start it is worth noting that your finished diagram should look something like the one in figure 10, with more or less shade obstacles depending on your site.

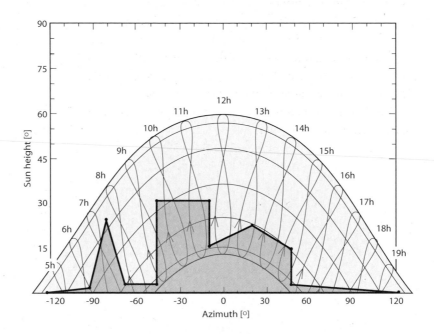

Fig. 10. Sun-path diagram with shading included.
Image courtesy of PV-Syst software, Dr Andre Marmoud, University of Geneva.

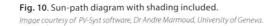

Finding site specific longitude, latitude and sun-path diagrams on the internet The diagram on page 16 (figure 9) is for a town at latitude 53.6°N and longitude 3.8°W, but diagrams can be generated for any location from a number of websites which you can find by searching for "Generate sun-path diagram" with an internet search engine. An example would be http://solardat.uoregon.edu/PolarSunChartProgram.html Here you can type in your longitude, latitude and time zone and download the resulting sun-path diagram as a PDF, which you can then print out and start drawing on.

To find out the exact longitude and latitude of your planned location you can use online mapping software at www.satsig. net/maps/lat-long-finder.htm Just double click on the world map and keep zooming in until you get close enough to read the figures for your home town, including the time zone.

Once you've printed out your sun-path diagram you can start plotting your shade objects.

How to map shade objects on your sun-path diagram

1) Stand at a location as close as possible to the site of the proposed PV system. Obviously you may not be able to stand at the height of a roof mounted PV array, in which case you will need to compensate for this in your readings (see technical box on page 26).

2) Take the first azimuth reading for a bearing of 60°. Remember – to get the azimuth reading, you need to subtract 180° from your compass reading, so at this bearing the azimuth reading is – 120°.

Fig. 11a

Hold the compass as shown in figure 11a with the dial end towards you. Rotate the dial on the compass until it reads 60° (against the marker at the top).

3) Holding the compass flat and keeping it pointing in front of you, turn your body until the red North end of the needle lines up exactly with the red North arrow on the dial behind it (figures 11b and 11c). You are now facing the bearing of 60° (azimuth -120°).

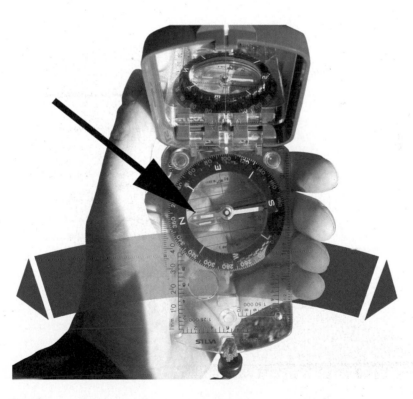

Fig. 11b

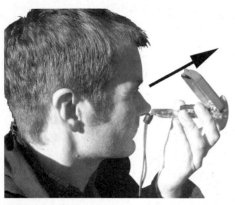

Fig. 11c

Fig. 12a

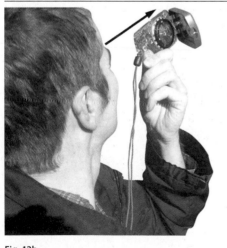

Fig. 12b

4) Hold the compass up to your eye and look along it (like a gun sight) to see what object is at the highest point on the horizon directly in front of you. This is easier to do accurately with a sighting compass which will have a mirror or prism to help view the needle while you look at the object, but it can be done with reasonable accuracy and a bit of practise using a simple hiking or orienteering compass.

5) Fine tune the compass by moving left and right so the needle is exactly parallel with the arrow on the dial behind.

6) Make a mental note of the object that is at the exact point on the horizon that you are facing.

7) Now use an inclinometer to work out the height of the object. Look along the inclinometer to line it up with the highest point on the horizon you just found with the compass.

Fig. 11c. Select an object on the horizon to use as a plot point, which can be aligned looking along the compass sight, whilst keeping the compass needle aligned with dial.

Fig. 12a. Look along the inclinometer to line it up with the selected object, then take a reading by pressing the hold button.

Fig. 12b. Some compasses have a built in inclinometer. To use it the compass is turned on its side. Look along the edge of the compass to line it up with the highest point on the shading object. The inclination is shown by a hanging pointer inside the compass dial.

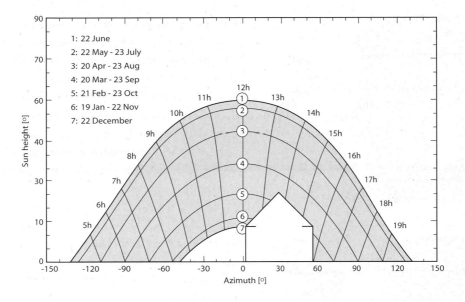

Fig. 13. Sun-path diagram with shading object plotted on. Solar paths at Dublin, (Lat.53.6°N, long.3.8°W, alt. 100m)

Image courtesy of Max Fordham and Partners (1999) 'Photovoltaics in Buildings: A Design Guide' DTI URN: 99/1274.

8) Take a reading with the inclinometer. On an electronic inclinometer the reading is taken by pressing the 'hold' button. On a sighting compass with built-in inclinometer, the reading can be viewed in the mirror.

9) Mark the inclinometer reading on the sun-path diagram according to the *x* axis on the sun-path diagram at -120° azimuth.

Repeat this process for different points along the horizon, and finally join up the points with a line, which should look a bit like your horizon. If it is a complex urban horizon you will need lots of points for an accurate representation, whereas a very flat horizon in a rural area will have fewer.

Once the shading is plotted onto a sun-path diagram, you can then see at what times of the day the PV array will be shaded for different times of the year. For example, the sun-path diagram on page 23 (figure 13) has a house plotted onto it. From the radial lines we can see that the house would shade the PV array in the afternoon. However, from the sun-path curves 5, 6 and 7 and the key, we can see that the house will only shade our PV array from September to March at those times.

Note that shading only affects direct sunlight coming onto an array. In the UK, much of the solar irradiation received by a PV array will be diffuse irradiation (due to particles of dust, water and gases in the atmosphere – see Technical box 'Air mass' on page 14 for more details). Diffuse irradiation is received by a PV array from the entire sky, so is much less affected by shading.

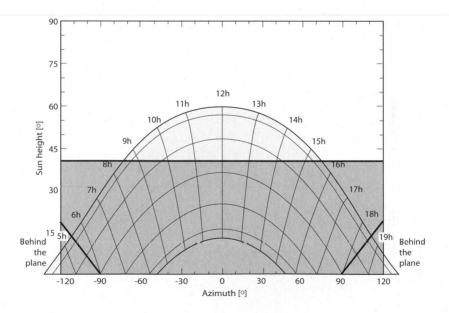

Fig. 14. Sun-path diagram with shading object (large warehouse) plotted on. Plane: tilt 30°, azimuth 0°.

Calculating the exact effect of shading on the performance of a PV system is very complex, but it can be done automatically using a computer-based PV system design programme, for example PV-Syst or PV*SOL (see Appendix 1: References and further reading). However, apart from the demo versions available for download, these systems are aimed at professional designers and are quite expensive to buy. Obviously, the impact of any localised shading objects will be less if your PV array can be located high up near the ridge of a roof, rather than as a free standing array in a garden (though this may be your only option, on a listed building for instance).

For example, if we test the hypothetical impact of a large warehouse shading our system by effectively raising the horizon by 40°, this reduces the output of a typical system by about 40% (see figure 14 opposite).

The graph below shows the reduction in output for horizons up to 60° (at which point the array receives no direct light and performance is reduced by about 70%).

% of optimum output

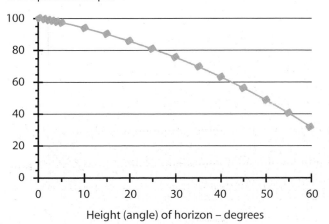

Height (angle) of horizon – degrees

Fig. 15. Reduction in output caused by shading.

TECHNICAL ——————————————————

Converting an elevation angle to a height

The tools

• Inclinometer

• Tape measure or measuring stick

• Scientific calculator

The calculation

WARNING!! This calculation uses trigonometry!

This is not essential information to create a sun-path diagram but you might find it helpful for other aspects of your installation.

It is possible to work out the height of an object (Y) if you know the distance to the object (X) and the elevation angle (E). The height can be calculated using a scientific calculator or computer. If you have an Apple computer you can download eCalc as a widget. Microsoft Windows has a built in calculator. To access it click start>programmes>accessories>calculator. There are also various scientific calculators on the internet to use free of charge – like the one at:
www.calculator.com/calcs/calc_sci.html or http://my.hrw.com/math06_07/nsmedia/tools/Sci_Calculator/Sci_Calculator.html

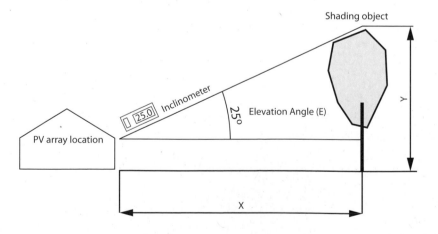

Fig. 16. Converting an elevation angle to find the height of an object.

The calculation below helps you to understand the maths. If you have internet access you can also get the answers by going to www.google.com and typing the figures into their search facility – like this: 15 meters*cos (30 degrees). The answer (12.99) comes up within seconds.

Y (height) = X (distance) cos E (inclination)

For example, if the distance is 15m and the inclination is 30°:

Y = 15 x cos 30 = 12.99m

For example if the distance X is 15m, on a calculator type:

The screen should display

NB. These calculator instructions are for the older type of calculator which requires each operation to be done separately. Newer calculators with alphanumeric displays require the formula to be entered in the format of the original formula:

which will be displayed on the screen as:

| 1 | 5 | COS | 3 | 0 |

Until the equals button is pressed

| = |

and the answer shown is:

Converting a known object height to an elevation angle

The calculation

Elevation angle $E = \cos^{-1}(Y/X)$

$$E = \cos^{-1}(13/15)$$
$$E = 30$$

Orientation and tilt elevation

Using tracking devices to optimise the angle of orientation

Ideally a PV array should be angled directly towards the sun at all times, so that less light is reflected off the cells and glass cover and more is absorbed by the solar cells.

One way to optimise the tilt and orientation of a PV module is to use an automatic tracking system that tilts the array in one (east-west) or two directions (east-west and up-down) and follows the sun throughout the day, as shown below.

When installed correctly, these trackers can improve PV output by up to 20%. However, they do require the solar array to be free-standing on the ground or on a flat roof. This will not be practical for many urban sites. Tracking systems also add

Fig. 17. Polycrystalline array with single axis tracking – West Beacon Farm, Leicestershire.

moving parts to the PV system, such as motors and bearings. These require maintenance, a disadvantage considering PV systems otherwise require little maintenance. As the cost of maintaining technical equipment can be nearly as costly as buying new, designing systems for low maintenance is a high priority.

Furthermore, if a tracking device breaks down in the wrong position, the PV system will perform worse than if there had been no tracking at all. So, whilst PV tracking has its place – for example, where a pole mounted system is more appropriate due to space constraints – for most systems the extra cost of tracking may be better spent by fitting more modules and a larger inverter to increase output (see page 56 for more on inverters). If you are considering a tracker system, compare different brands and ask suppliers how long their trackers have been in production. Also ask where you can go to see working trackers that have been in use for several years.

Optimal tilt depends on latitude

The ideal roof for a fixed PV array in Britain and Ireland would face due south and have a tilt of 30° to 40° depending on latitude. The exact optimum angle of tilt is approximately the latitude of the site minus 20°. For example, London has a latitude of 51° so the perfect pitch would be 31°, but in Edinburgh – with a latitude of 53° – the perfect pitch would be 33°. The northernmost settlement in the Shetland Islands is Haroldswick, with a latitude of 60°. As Land's End in Cornwall has a latitude of 50°, the latitude range for Britain is 50°(Land's End) to 60°(Haroldswick). Hence, the optimal tilt angle is between 30° and 40°.

For latitudes beyond the British Isles, the optimum angle of tilt can be found using a computer model like PV-GIS or PV-Syst or by checking your latitude at www.satsig.net/maps/lat-long-finder.htm and deducting 20 from the figure you find.

Fig. 18. Various slopes and pitches on traditional and modern British roofs and energy losses caused by pitch variations.

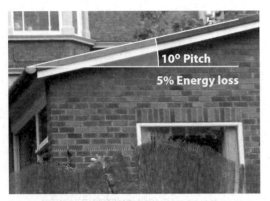

Seventies detached house with almost flat roof, (hence roofed with felt & not tiles) 10° pitch 5% energy loss.

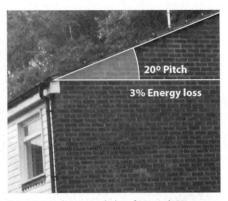

Seventies end terrace, tiled roof 20° pitch 3% energy loss.

Nineties detached house, 24° pitch 2.5% energy loss.

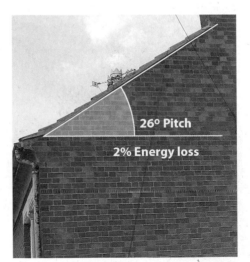

Victorian terrace, 26° pitch 2% energy loss.

Fig. 18 continued

Thirties semi, 26° pitch 2% energy loss.

Victorian detached, 43° pitch 2.5% energy loss.

Victorian detached 48° pitch 3% energy loss.

Mansard Dutch-style roof, 88° on steeper pitch,
almost a facade, 88° pitch 30% energy loss.

How tilt can affect performance in solar PV systems

Don't be put off if your roof turns out to have a less than optimal pitch. Most roofs with a tilt angle of between 10° and 50° will be above 97% of the optimum. Nevertheless, when designing a system it is important to know how much power you are likely to lose due to your tilt angle – especially if you are claiming feed-in tariff incentives and selling some of your power back to the grid. Over a 20-year period a 3% loss of power can add up to a significant amount.

The photographs in figure 18 on the previous pages show the effect of typical roof pitches on performance, for homes in England, Wales and Ireland with a latitude of 52°.

Measuring the roof pitch

You can measure the pitch of a roof accurately using an inclinometer, shown below.

You may also be able to measure the pitch using a spirit level. Some types of spirit level have an angle finder which can be used for measuring pitch as shown in figure 20.

Fig. 19. Measuring roof pitch with an electronic inclinometer

Fig. 20. Measuring roof pitch with a spirit level incorporating an angle finder.

For a high roof, the pitch can also be measured from the ground by sighting along the inclinometer as shown in the sun-path diagram shade plotting example earlier in this chapter, but note; whilst this method is accurate enough for system yield analysis, it wouldn't be accurate enough for the design of roof modifications (see Chapter Seven).

Roof orientation (azimuth)

PV arrays in the northern hemisphere would ideally face due south (due north for the southern hemisphere). This is most important in more northerly latitudes where the sun is generally lower in the sky. However, space constraints in urban housing estate design mean that few roofs will face due south.

The azimuth angle

To visualise what the azimuth angle actually represents it might be worth doing a little experiment using a compass, two lines of string or a piece of chalk. Stand on a solid piece of ground and find due south with the compass. Leave an item of clothing on the floor where you were standing (to represent your solar PV array) and draw the string or a chalk line out due south from the clothes for a couple of metres.

Now go back to the clothes and look towards (though not at!) the sun. Draw a line out from the clothes a couple of metres in the direction of the sun. The azimuth angle is the difference between the due south line and the sun line. If you imagine your clothes as the solar array and draw out lines throughout the day you can get a pretty good visual clue as to when your solar panel will be performing at its best. The azimuth angle will change throughout the day as the sun moves around, and thus so will the efficiency of your PV module.

The diagram below shows the effect of roof orientation (azimuth) on solar PV array performance for homes in England, Wales and Ireland with a latitude of 52°.

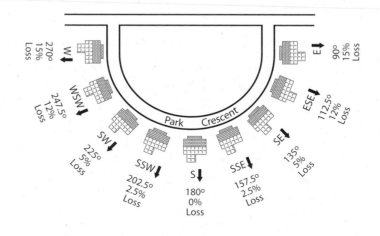

Fig. 21. Diagram showing the effect of array azimuth (angle of orientation) on performance for 52° latitude (PV arrays on roof pitches facing away from street). *Image courtesy of Max Fordham & Partners (1999) 'Photovoltaics in Buildings: A Design Guide' DTI URN: 99/1274.*

The figures in this diagram are based on arrays facing away from the street. It is clear that roofs facing south-east to south-west will perform at up to 95% of optimal output, whereas east or west facing roofs reduce performance by 15%. This reduction in output isn't a major technical impediment, it can be compensated for by installing a slightly larger array which will generate the same amount of energy as a smaller array at the optimum orientation. (Note that these figures assume all other factors are perfect, that is, zero shading, good ventilation and so on).

Figures 21 and 22 show the effect of tilt and orientation and give an approximate guide to performance. To establish the combined effect of tilt and orientation on performance more accurately, a polar graph for the exact latitude is used. A typical graph for latitude 52° is shown below. For example, for a roof facing south-east with a 50° angle of tilt the performance result would be 10% below the optimum performance of a 35° roof facing due south.

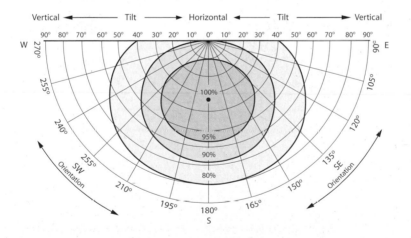

Fig. 22. Graph showing effect of tilt and azimuth on irradiation on an array in Cambridge (52° latitude), the optimum tilt shown for optimum output (100% point) is 32°.
Image courtesy of Max Fordham & Partners (1999) 'Photovoltaics in Buildings: A Design Guide' DTI URN: 99/1274 where other graphs can also be found (published by BERR – Department for Business, Enterprise and Regulatory Reform).

Fig. 23. Measuring the orientation of a building with a compass.

Checking the orientation of a roof is relatively easy (see figure 23), since the side walls of a building are usually parallel with the orientation of the roof. One can simply hold a compass against the side of the building and take a bearing (then subtract 180° from the reading to convert the compass bearing to an azimuth angle).

Conclusion

From this chapter you will have an idea of the ideal conditions for a PV system – south facing, 35° pitch and no shading – so if your only available roof is north facing, or is heavily shaded by a tower block to the south, then you might like to rethink at this point. You will also begin to appreciate the complexity of the many factors affecting PV system performance, which is where the various software packages and websites listed are helpful as all the factors can be taken in conjunction.

Chapter Three

The photovoltaic effect

Modules and cells

As mentioned in the last chapter, the term 'solar panel' can be misleading because it could refer to a solar thermal collector used for heating water, or a PV module used for generating electricity. Here, to be more precise, the term 'module' is used. A solar PV module is made up of a number of individual solar PV cells, which actually do the business of generating the electricity.

Fig. 24. A single PV cell.

Fig. 25. A PV module made up of multiple cells.
Image courtesy of Dulas Ltd.

The cells themselves are fragile and only generate a tiny amount of electricity. They are connected together to increase the power output and encapsulated in a weatherproof laminated module consisting of a glass front, aluminium frame and protective back cover.

Each solar module has a power rating which tells you how much electrical power the module will produce under ideal solar irradiance conditions (also described in the last chapter). Some very small stand-alone systems – such as parking meters and street lights – only require a single module, but usually two or more modules are connected together to ensure enough energy is collected. A group of PV modules connected together is called a solar PV array.

How solar PV cells work

A short history of the PV cell

Alexander Edmond Becquerel first came across the photovoltaic effect in 1839 as a side effect within his wider photochemical research. Following on from this early discovery, in 1873, John Willoughby Smith published the results of his experiments with the photoelectric effect when examining the electrical properties of selenium used for testing undersea electric cables. He noticed that the selenium's resistance to electric current was reduced when it was better illuminated. In 1876, W. G. Adams and R. E. Day at Cambridge University researched the solar effect on selenium and were able to create a current through the selenium using light from a candle, with no other energy source present, describing it as the 'photoelectric' effect.

By 1883, Charles Fritts, a New York electrician, had developed the selenium cell further, sandwiching layers of selenium between a metal backing and transparent gold leaf cover. He called the device a selenium battery and achieved an efficiency of less than 1%. Fritts was unable to gain recognition for his device because the engineering establishment didn't think it possible to generate electricity without consuming matter, as in a combustion engine. Fritts sent a model to Werner Siemens, an established German inventor and industrialist, who was able to replicate the effect and published the outcome.

In 1954, Calvin Fuller and Gerald Pearson of Bell laboratories were experimenting with silicon rectifiers and found that the properties of the device varied depending on where the device was positioned in the lab. Having realised the difference was due to lighting, they recorded a maximum efficiency of 4%, five times that of the much earlier selenium device. Fuller and Pearson joined forces with Darryl Chapin, who was working for Bell on possible power sources for remote telecommunications repeater stations. Together the team managed to improve the efficiency of their silicon device to 6% and displayed a solar-powered model Ferris wheel and FM radio at the National Academy of Science in Washington DC. *The New York Times* reported: 'Vast Power of the Sun is Tapped by Battery Using Sand Ingredient'. The team improved their lab efficiency to 15%, but were realistic that the technology was still at a costly experimental stage. The project was eventually dropped by Bell in favour of cheaper conventional energy sources, but the company licensed the technology to others, including one that marketed PV-powered forest look-out towers and coastguard buoys. However, US government agencies thought the future of remote power supplies would be in small nuclear reactor-based devices rather than PV.

Ultimately, PV technology was largely rescued by the US space programme, which required a lightweight, compact, autonomous energy source for its satellites. PV was perfect for its requirements and from the late 1950s solar cells powered all US satellites. This ensured that at least some research continued, developing PV technology in a world that at the time was focusing on nuclear and fossil fuel energy sources for most applications.

Since then, research and development has continued, increasing PV cell efficiency, and PV is now installed in sites all over the world.

Buying the module with the right kind of cell

There are a variety of materials that can be used to convert light directly into electricity as part of a solar cell. Different types of solar cell have varying efficiencies and costs. When choosing a solar PV system it is important to have some knowledge of the main technologies and their relative merits and deficiencies so you can choose the appropriate product for your project.

Comparing PV cell types – which cell type do I need?

The primary conductive material used by 95% of all solar cells is silicon. Silicon is the second most abundant element on the planet after carbon. There are two types of silicon solar cell: *crystalline* and *amorphous* (a type of thin film). Crystalline cells are cut from a solid ingot of silicon, whereas amorphous cells are made by depositing a silicon semiconductor compound onto a sheet of glass.

Crystalline silicon has higher efficiency and better longevity but requires more raw material than amorphous silicon and is thus more expensive to produce. It is the popular choice for solar PV systems on homes and offices where the module must withstand the weather and other environmental factors, and where the cost of accessing roofs to replace failed solar cell modules is an important factor in determining the financial viability of a project.

Amorphous silicon cells use much less raw material and are cheaper to produce than crystalline silicon, but are less efficient and suffer more easily from degradation. They tend to be used for relatively short-lived consumer appliances, such as calculators, garden lights and portable battery chargers.

Crystalline solar cells

The two main types of crystalline silicon are monocrystalline (mSi) and polycrystalline (pSi). Monocrystalline silicon is more expensive and more efficient than polycrystalline silicon. Polycrystalline has a blue sparkly appearance whereas monocrystalline has a dark grey homogeneous appearance; sometimes the choice between the two is an aesthetic one, to achieve either an eye catching or discreet appearance to match adjacent roof or wall coverings.

Gallium compounds, such as gallium arsenide (GaAs), are also used for crystalline solar cells. These can achieve higher efficiencies but are much more expensive to produce and so are normally used for satellites and space stations where the efficiency is critical and the cost can be justified.

Fig. 26. Polycrystalline silicon solar cell.

Fig. 27. Monocrystalline silicon solar cell.

Thin film solar cells

Materials used for thin film solar cells include amorphous silicon (aSi), cadmium telluride (CdTe) and cadmium indium diselenide (CIS).

Amorphous silicon (aSi) modules are also available as flexible modules, which can be used on flat or curved surfaces. These can be supplied with eyelets for mounting on yachts or caravans, for example. ASi modules can also be thermally welded onto modern flat roof materials making it possible to install PV modules when a roof is laid.

Using solar modules bonded to the roof incurs a slight performance reduction in northerly latitudes because they will not be placed with the optimum tilt towards the sun. However, they do have the advantage of avoiding the wind and snow load issues common to framed flat roof systems which might otherwise prohibit the use of PV modules on lightweight flat roof structures (see Chapter Seven).

Fig. 28. Amorphous silicon solar modules.

Fig. 29. ASi module welded to flat roof.
Image courtesy of Solar Integrated Technologies GmbH.

CdTe and CIS cells are now available to install, but the efficiency of manufactured modules is still significantly less than that achieved in laboratories. However, it is expected that higher efficiency modules will be available in the future.

Tandem junctions

Some manufacturers now use multiple layers of solar cells in order to capture more of the sun's energy. These are called tandem junctions. Twin and triple junctions are now commonly available, and greater numbers of layers may be available in the future. Often the layers are different types of solar cell sandwiched together, as different cells respond better to different colours within the spectrum of light in the sun's rays. When we talk about this solar spectrum we are referring to the colours that normally appear as mixed together in the white light from the sun, but which can be separated with a prism or in a rainbow.

Fig. 30. CSI thin film module.

Environmental concerns

There are environmental concerns about the use of toxic compounds used in certain brands of PV module, for example cadmium. Finished PV modules pose little health risk because the cadmium compound is encased in a weather-tight glass laminate and is in a stable form. Cadmium is produced as a by-product of zinc mining in developing countries, where there are concerns about pollution and its effects on the local population. However, the amount of cadmium used in PV manufacture is fractional compared to that in NiCd batteries which are exempt from the 'Restriction of Hazardous Substances' directive (RoHS). Cadmium is also released from coal when it is burnt in power stations, and in greater quantities per kWh generated than released in producing PV cells capable of generating the same number of kWh.

Understanding amps, current and voltage

With regard to electrical terminology, up until now it's only been important to know what's what about watts. From here on in you'll need at least some familiarity with the terms 'amps' and 'volts'. These will help you understand what is happening to the electricity your PV system produces as it passes from the modules through the wires and components and into your home. Electricity moves around wires in different strengths and at different speeds, and the strength and speed determines the types of components and appliances you need to buy. It also dictates the safety measures you have to take and the efficiency of your overall system.

Amps and current

When electricity travels through cabling to your socket it does so at a particular rate of flow. This flow of electricity is described as the 'current' and is measured in amps, which is short for ampere and has the symbol 'A'. At some point most of us have changed the fuse in a plug, so we are vaguely familiar with the term amp. The fuse limits the flow of electricity delivered via your electrical socket to a particular appliance in order to prevent a fire if there is a fault.

It is important not to confuse amps with 'amp-hours'. In the context of stand-alone systems (those not connected to the grid), amp-hours describe battery capacity – ampere multiplied by hours, not amp per hour. For example, a 200 amp-hour battery will give 20 amps of current for 10 hours, or 10 amps of current for 20 hours (see page 59).

Electricity can be controlled and directed to flow in different ways and so we also talk about two different types of current – 'Alternating Current' and 'Direct Current'. Alternating Current (AC) describes the type of electricity in the mains supply used for household appliances. Batteries and solar modules produce Direct Current (DC), which is also used in vehicles and rechargeable or battery-powered appliances.

Volts and voltage

All electricity must be 'pushed' through cabling to reach its target; driven from A to B. This driving force is measured in volts (V) and is described as voltage. Historically, PV modules were manufactured with a nominal output voltage of around 16-18V, which is ideal for charging a 12V battery. Voltage can be varied by both system design and electronic controllers to provide the necessary 'push' required for a particular system. For example, if you require a 24V system you would need to connect two PV modules and two 12V batteries in series (as shown in figures 35 and 37 on pages 57 and 59).

TECHNICAL ————————————————————

Crystalline solar cell manufacture

The cell itself is fabricated by slicing a thin wafer of PV grade silicon off an ingot using a wire saw. Chemical processes are then used to build up negative and positive charged layers of silicon in the form of a wafer. It is at the boundary between these layers – known as the depletion zone – that the incoming photons of light cause electrons to move from the negative charged side of the layers to the positive charged side. The negative front surface of the cell is etched with a fine jagged pyramidal surface, which makes it better at absorbing light.

A fine lattice of silver contacts is printed onto the front of the negative layer onto which electrical connections can be soldered. The branches of the lattice are kept as thin as possible to maximise absorption of light by the negative layer. The negative layer is very thin (approximately one thousandth of a millimetre) so light is readily transmitted through it. The positive layer is comparatively thick (approximately one third of a millimetre) to prevent breakage during production. A continuous metal foil is sprayed on the back of the positive layer onto which electrical connections can also be soldered. The metal foil is continuous across this surface because light transmission isn't an issue at the back of the cell and It allows maximum current to flow.

Negative front contact arranged in thin strips to minimise shading

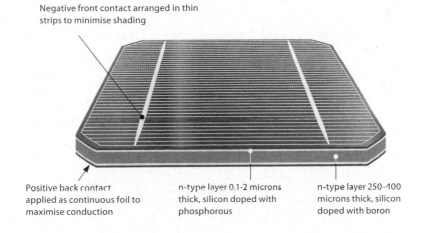

Positive back contact applied as continuous foil to maximise conduction

n-type layer 0.1-2 microns thick, silicon doped with phosphorous

n-type layer 250-400 microns thick, silicon doped with boron

Fig. 31. Cutaway section of crystalline PV cell (not to scale).

Each cell generates about 0.5V, which is of limited practical use, so 36 cells are usually connected together in series strings to give a nominal module voltage of 18V. Modules were historically supplied with a nominal voltage of 18V because most PV modules were used to charge 12V batteries. The voltage of any source of electricity must be higher than battery voltage to overcome cable losses and allow current to flow – for example, an alternator typically generates 14-15V at full speed when charging a 12V car battery.

The in-series solar cells are then laid onto a sheet of high quality glass, encased in EVA (a soft sealant with similar properties to silicone sealant) and covered with a UV-resistant Tedlar™ plastic film at the back. A junction box is bonded to the back with either terminal blocks inside or trailing leads with weatherproof connectors.

The semi-transparent PV modules used to glaze atrium roofs simply have a second sheet of glass behind the cells instead of Tedlar™, though this makes them considerably heavier. Many roofs would not be strong enough to support this type of module.

The glass/PV laminate produced part way through this process can often be supplied for use in low-profile roof-integrated arrays as shown in the diagram below, but is more usually placed in an aluminium frame for mechanical strength. The aluminium frame is a specific extrusion designed to bolt quickly onto PV mounting rails.

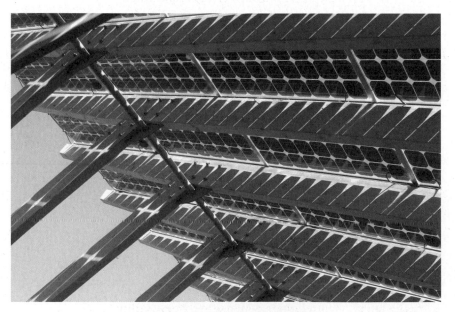

Fig. 32. The new PV roof on CAT's WISE building.

Chapter Four

Stand-alone or grid-connected?

Introduction

If having read the preceding chapters solar PV still appears to be an appropriate energy choice for your situation, it is important to be clear about whether the solar PV system will be linked to a utility electricity supply ('grid-connected') or be stand-alone (off-grid), as these two system types use different parts and are designed differently.

Storing power

The intensity of light reaching a solar PV array fluctuates considerably due to the factors highlighted in Chapter Three. Variations in the solar resource are even more dramatic than the human eye perceives them to be. This is because the irises in our eyes expand and contract to compensate for different levels of light.

If you were to connect a light bulb directly to solar PV without a battery (or utility connection) in between, it would flicker as clouds pass in front of the sun, varying irradiance, which would be very irritating. Therefore, any solar electricity system requires some means of smoothing out these variations. Stand-alone systems use storage to achieve this, usually in the form of batteries. A battery is charged when there is lots of sunlight and discharges when there is less. Alternative storage solutions such as hydrogen or flywheels are being researched but are currently a long way from being efficient and cost effective.

Grid-connected systems avoid the need for storage because electricity can be imported and exported to and from a utility company's electricity network, which acts like a 'virtual battery' (discussed below).

Lead acid batteries are the common choice for stand-alone power systems because they are currently the best compromise between cost, efficiency and longevity. However, they have a limited lifespan, are environmentally problematic, require careful maintenance and add to the capital and running costs of a power system.

In buildings that already have a utility electricity supply, a grid-connected renewable energy system is significantly cheaper to install than a stand-alone system because it avoids the extra capital and maintenance costs of installing batteries.

Connecting to the utility electricity network

Historically, a regulatory framework for grid-connected renewable energy systems did not exist and so suitable systems to connect safely to the grid were not available. The only way to power your home with renewable energy was to have a stand-alone system, operating completely separately from the utility company supply.

The decision to choose stand-alone power was often also environmentally and politically motivated. Until very recently, buying domestic electricity meant buying into coal, oil, gas and nuclear power generation. Now that green energy tariffs are readily available the environmental arguments against connecting to the grid are very weak, for reasons which will become apparent later in this chapter.

If you have a stand-alone renewable energy system, you need to be able to guarantee sufficient generation and storage capacity to see you through times when your energy resource is low but your demand is high. Hence, you will have to buy a PV array with a greater capacity than your average demand and install adequate storage to provide enough electricity at times of low generation. This has serious implications for the cost of a system.

Conversely, the major advantage of grid-connection for renewable energy systems is that of 'aggregation'. Individual consumers and businesses connected to a utility system (or 'mains') have a highly variable rate of use of power depending on their activities. However, because domestic, commercial and industrial users of electricity are connected to the same utility system – the national grid – the total demand is averaged out both regionally and nationally resulting in much less variation overall. This means that less generating capacity is required to supply all users at all times than if each had their own generating system.

This aggregating effect works equally for renewable energy systems on the grid; whilst one solar array may be under cloud, or one wind farm may experience a calm period, by grid-connecting all the renewables systems together, the output is aggregated and variations in resource have less of an overall impact on the network.

The national grid has spare generating capacity to cope with sudden demand surges (for example, when everyone boils a kettle during popular TV programme breaks, or if a part of the system experiences a fault). This surplus capacity, which is required to maintain reliability during faults, also allows for spare capacity during calm periods when the output from wind farms is reduced. It is often suggested in the media that we therefore need an equal amount of spare gas/coal generating capacity for each wind turbine and solar PV array. This is like saying every fire station must use a different fire engine to attend road accidents from the one it uses to attend fires! They won't all be experiencing calms or cloud at the same time.

So, to summarise the advice of most renewable energy consultants; if you already have a grid connection or are near enough to a ultility supply that grid connection isn't prohibitively expensive, then a grid-connected system will be more reliable, require less maintenance and be cheaper to run per kWh. However, if you are in a very remote location, a stand-alone renewable energy system will be much cheaper to run long-term than a petrol or diesel generator, which might cost 50-60p/kWh.

Feed-in tariffs

Feed-in tariffs reward generators for the amount of renewable energy they produce, whether they are connected to the grid or not. There is an additional tariff available for those who do export to the grid but at 3p per unit it is much smaller than the main tariff (up to 41.3p). This arrangement recognises the contribution stand-alone renewable energy systems make to reducing carbon emissions in remote areas from generators running on petrol and diesel.

Chapter Five

Stand-alone PV systems

Introduction

If there are no power lines within reasonable distance of a proposed solar PV system then the cost of getting a utility electricity connection may mean a stand-alone renewable energy system is the only option. This may also be true in countries where the connection of small-scale renewable systems to the utility network is difficult, perhaps because a system of regulation and approvals is not yet in place.

Stand-alone PV systems are used for a wide variety of applications such as: lighthouses, marine marker buoys, parking meters, traffic and weather monitoring, street-lighting, garden lights, vehicle trickle chargers, yachts, caravans, mobile homes, fountains, calculators, water-pumping and vaccine fridges.

Energy storage will be necessary for the majority of stand-alone applications, usually in the form of batteries. In the future, fuel cells might be used for stand-alone systems, but at present they are costly and inefficient in comparison. For very small-scale applications, such as caravans and mobile homes, storage will usually require a 12V battery, 24V for larger systems (a house for example). For larger commercial stand-alone systems a 48V battery store is sometimes used.

What you will need for a stand-alone system

Batteries

In some countries, car and truck batteries are used for PV systems because they are cheap and readily available. However, this is a false economy; vehicle starting batteries are designed to supply a short burst of high current to start the car and then be trickle-charged by the alternator the rest of the time. In any stand-alone power system (PV, wind or hydro) batteries will be deeply discharged and sometimes left discharged for long periods; vehicle starting (or 'cranking') batteries aren't designed for this and will permanently lose their ability to store electricity even after a brief exposure to this kind of use.

Batteries used in stand-alone energy systems must be deep-cycle batteries; these can be leisure batteries designed for yachts and caravans, or second-hand forklift, submarine and uninterruptible power supply (UPS) batteries. However, finding surplus dealers who can supply batteries which still have some working life can be difficult. Ideally, batteries specifically designed for stand-alone renewable energy systems should be used as they will last longer. These are available from renewable energy equipment distributors.

Batteries are surprisingly sensitive given their robust external appearance; they must be specified, installed, operated and maintained carefully if they are to perform reliably for their expected lifetime.

Having chosen the correct battery, the system should be designed for maximum battery life, by using load and charge controllers to limit the amount of charge and discharge.

Load controller

Here the word 'load' refers to the demand that will be placed upon your electrical system by the equipment you use in your home: televisions, fridges, computers, washing machines, and so on. For a stand-alone system the load defines the number of PV modules you will choose to purchase for your array (see also the sizing section at the end of this chapter). A load controller is a device that electronically disconnects appliances and/or lighting when batteries are fully discharged (in other words, flat) to prevent battery damage.

Charge controller

Conversely, a charge controller is a device which limits battery charging when batteries are fully charged, again to prevent damage. Sometimes the charge and load controller are combined in one unit with other functions (for example, a display showing voltage and current). The charge controller prevents overcharging by electronically disconnecting a PV array when a battery is fully charged. The charge controller is connected to the system between the array and the battery.

Maximum Power Point Tracker (MPPT)

Maximum Power Point Trackers (MPPT) are available for stand-alone systems. These optimise the voltage and current of the PV array to achieve maximum output from the solar modules. They also incorporate a charge controller.

Fig. 33 (below right). Maximum Power Point Tracker (incorporating a charge controller).

Fig. 34 (below left). A combined charge and load controller.

Inverter

All power systems are labelled according to their voltage. Many stand-alone systems are operated as 12V or 24V DC (Direct Current) systems. There are appliances that run off 12V and 24V DC systems but the vast majority of household equipment is designed to be used with a standard 230V AC domestic power supply (except Japan and North and Central America which use 100V, 110V and 120V supplies). You can still use normal appliances in a DC system but before doing so you have to pass the electricity through an inverter to convert the voltage from DC to AC (see page 83 for more details on inverters).

Fitting the different parts together

Understanding series and parallel circuit connections

Most solar PV systems rely on batteries to supply an even flow of electricity. The number of batteries needed depends on final load calculations (see page 66), the storage capacity of the battery (labelled as an amp-hour (Ah) rating) and the amount of power required to be ready for use at any one time. Systems requiring only very small amounts of power might only need one battery, but most require two or more. The modules and batteries may be arranged in series or parallel according to whether a 12V, 24V or 48V system is required.

It is normal for a power source (in other words your PV module) to have a higher voltage than the battery it is charging; a car alternator for example produces up to 15V but charges a battery with a nominal voltage of 12V. Some PV modules are manufactured with a nominal output voltage of around 16-18V, which is ideal for charging a 12V battery. But if you have a 24V system, you would need to connect two PV modules and two 12V batteries in series as shown in figure 35 opposite and figure 37 on page 59. A series connection allows

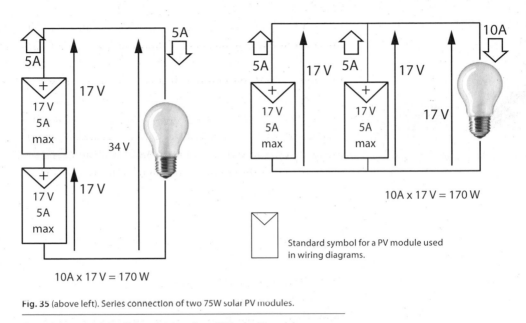

10A x 17 V = 170 W

Standard symbol for a PV module used in wiring diagrams.

10A x 17 V = 170 W

Fig. 35 (above left). Series connection of two 75W solar PV modules.

Fig. 36 (above right). Parallel connection of two 75W solar PV modules.

the voltage to be doubled. In figure 35 two solar 17V PV modules have been connected in series to create a circuit with an overall nominal voltage of 34V. 24V and even 48V systems are used for larger stand-alone systems with long cable runs, as the voltage drop in the cables would cause dimming of lights and damage to electronic equipment.

A parallel connection delivers a lower total voltage but can deliver more current than a series system with the same number of components. In figure 36 above two 17V PV modules have been connected in parallel to create a parallel circuit. Despite having two modules, each with a nominal voltage of 17V, the total nominal voltage remains at 17V. It does not double to 34V as in a series circuit. This would not provide enough voltage to run appliances on a 24V system but could be used with a 12V battery.

You will notice that in the series circuit the current rating is the same as the current rating of the individual modules, in this case 5A. However, in a parallel circuit the current is the sum of the module currents, in other words, 5A+5A=10A.

SAFETY

Beware: if you want to have these circuits installed you also need some additional protection components, so don't attempt to set up a battery circuit from this description alone. It is just an illustrative example to show the fundamentals. Note also that the series connection works for DC sources, like batteries and PV modules, but would not be safe with AC sources, like inverter outputs (unless the inverters have synchronized communication to permit parallel connection).

Calculating total power in series and parallel circuits

In both the series and parallel circuits pictured on the previous page there are two 75W modules making the total power always 2 x 75W = 150W, regardless of whether the modules are wired in series or parallel. This is because in any circuit:

Power (P (Watts)) = Voltage (V (Volts)) x Current (I (Amps))

This stands providing we use the total voltage and current ratings of the circuit.

The same principles of calculating voltage, current and power apply to series and parallel circuits of batteries.

The circuit diagrams opposite (on page 59) look similar to those on page 57, but now represent battery wiring rather than module wiring. You'll note that the amp figures have been replaced by amp-hour (Ah) figures.

As mentioned in Chapter Three, the amp-hour unit relates to the amount of energy which can be stored by a battery and is usually printed on the battery label, for example, 110Ah or 220Ah. One amp-hour literally means the battery can supply one amp for one hour. This is a nominal value; in practice batteries are not completely discharged on each cycle (and complete discharging causes premature ageing of the batteries).

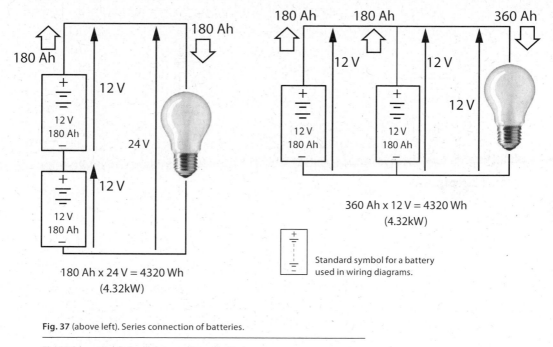

180 Ah x 24 V = 4320 Wh
(4.32kW)

360 Ah x 12 V = 4320 Wh
(4.32kW)

Standard symbol for a battery
used in wiring diagrams.

Fig. 37 (above left). Series connection of batteries.

Fig. 38 (above right). Parallel connection of batteries.

Converting amp-hours to kilowatt-hours

System load calculations are made in kilowatt-hours and so the results will not immediately help you to work out how many batteries you actually need. This is because battery power is rated in amp-hour units, not kilowatt-hour units, as mentioned above.

If you want to convert from Ah to kWh you'll need to use one of the following calculations, depending on whether you have a series or parallel set:

For a series set Total energy storage (kWh) = battery capacity (Ah) x battery total voltage (V) ÷ 1000

So using our series examples above:

kWh = 180Ah x (12V+12V) ÷ 1000

In other words…

kWh = 180Ah x 24V ÷ 1000

and breaking it down one more step …

kWh = 4320Wh ÷ 1000

which gives a final figure of…

4.32kWh

For the parallel circuit Use the same calculation but add up the two amp-hour figures rather than the two voltage figures. Total energy storage (kWh) = battery capacity (Ah) x battery total voltage (V) ÷ 1000

So using our parallel examples above:

kWh = (180Ah+180Ah) x 12V ÷ 1000

In other words…

kWh = 360Ah x 12V ÷ 1000

and breaking it down one more step…

kWh = 4320Wh ÷ 1000

which gives you a final figure of…

4.32 kWh

As you can see, the total energy stored is the same whether we connect the two batteries in series or parallel, the only thing that has changed is the voltage and current.

Running AC and DC appliances

It is possible to run both AC and DC appliances off the same solar system. In the case of 12V systems, there are appliances designed for cars, caravans and boats available that run directly off 12V including those that plug into a car lighter socket. For 24V systems there are some 24V DC appliances designed for trucks and coaches available, but it is also possible to buy 24V to 12V, and 12V to 24V, converters (although these do incur some efficiency losses; that is, wasted energy). Chargers for mobile phones and laptops that plug into a car lighter socket are now available that can also be used with 12V stand-alone renewable energy systems. An increasing selection of 12V and 24V DC low energy light fittings is also now available from renewable energy equipment distributors.

Regular domestic 230V AC appliances can be used on a stand-alone system but require an inverter to convert the electricity from DC to AC and step up the voltage. When we refer to AC equipment in a stand-alone system, we are generally talking about 230V AC (100/120V AC in Japan and Central and North America). Inverters range from simple 150W models – that can be plugged into a car lighter socket (which often have built-in appliance sockets where no AC wiring is required) – to 4kW inverters with built-in battery chargers. Whilst modern inverters are more efficient and reliable, they do add losses to a system, especially if left on all the time.

Many devices used in modern houses are only available for 230V AC (100/120V AC in Japan and Central/North America, as previously mentioned) – for example, washing machines and heating controls – so most stand-alone systems will need at least some 230V AC circuits running off an inverter.

The advantages of using DC equipment in a stand-alone system are simplicity and reliability. You also save some energy by not having a large inverter running on standby. However, using DC instead of AC appliances doesn't always cost less as DC appliances are often more expensive; also thicker (more expensive) cables, and hence larger fittings, are usually required for lower voltage DC circuits.

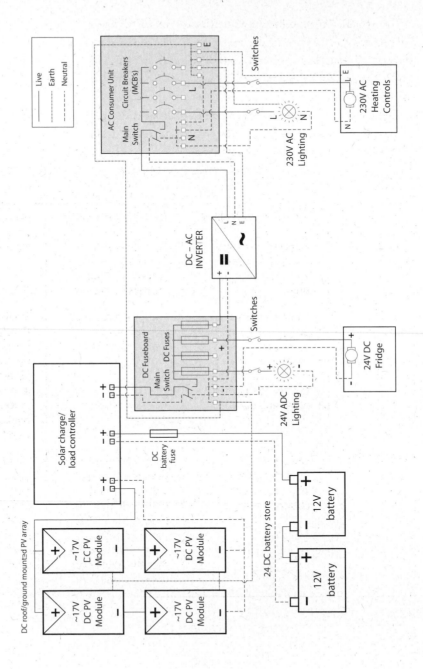

Fig. 39. Stand-alone PV system schematic.

The schematic opposite shows both AC and DC loads connected. You can see all the major components of a solar system and how they are connected using standard electrical cabling and other components, such as DC fuse boards and AC consumer units.

The circuit can operate in any one of nine modes depicted in the table below. The change from one mode of operation to another usually happens passively if the system is designed and installed correctly; no manual switching is required.

Appliances	PV array generating	Partially generating	Not generating
Not in use	Energy flows from PV array to battery and battery charges up.	Small energy flow from PV array to battery and battery charges up (slowly).	No power flows.
Partially in use	Some energy flows from PV array to battery and battery charges up. Some energy flows directly from array to loads, bypassing battery.	Energy flows directly from PV array to loads, bypassing battery.	Small energy flow from battery to loads: battery discharges slowly (unless the battery is fully discharged in which case the appliances will not work).
In use	Energy flows directly from PV array to loads, bypassing battery.	Energy flows from both the array and battery to loads: battery discharges slowly (unless the battery is fully discharged in which case the appliances will not work).	Energy flows from battery to loads: battery discharges (unless the battery is fully discharged in which case the appliances will not work).

Fig. 40. Energy flows in a stand-alone system.

Sizing a stand-alone PV system

Stand-alone PV systems require much more care when sizing than grid-connected systems because the system must meet all of the energy demand all of the time, even during extended cloudy periods. In contrast, a grid-connected system allows you to choose to supply only a proportion of your electricity needs as available finances might dictate.

An important criterion with a stand-alone PV system is the number of 'days of autonomy' for which the system must allow. If you require the system to run 24 hours a day, 365 days a year, you will need a much larger battery store to see you through any long periods of heavy cloud and rain, and a much larger PV array to keep the battery store charged; this approach is used commonly for things like vaccine fridges and navigation buoys, but is very costly. If you are prepared to accept some periods without electricity, your array can be much smaller (and hence much cheaper).

A popular approach to mitigate the problem of operating a stand-alone system through long periods of low irradiance is to have a 'hybrid' system consisting of a PV array and a wind turbine. The idea being that – in temperate climates – periods of low irradiance usually correspond with periods of high wind speeds and vice versa, both on a short-term and a seasonal basis. Hence, a hybrid system will produce energy at lower cost per kWh than a solar only system. This idea can be extended further to include a hydro turbine if you have a suitable site, or a diesel stand-by generator. Periodically running a diesel generator can mean a big saving on the size of battery store required and is thus a very common strategy for stand-alone systems.

Fig. 41. A hybrid system incorporating solar PV array and wind turbine.

Calculating how much electricity generating capacity you will need

Before a stand-alone renewable energy system can be installed the load must be calculated. This is the demand placed upon your energy system by all your appliances, lights and devices – in short, your electricity requirement.

The table below is an example of a simple excel spreadsheet which can be used to calculate how much a solar PV system would need to generate in order to supply your energy demands, although you can of course do this with a piece of paper and a calculator (the spreadsheet can be downloaded from www.briangoss.co.uk). This is a simplified version to show how the numbers work. Many more rows of information will usually need to be added.

To size a stand-alone system, the system designer (or you) will need to assess what appliances will be used and for how many hours per week.

Appliance description	Number of appliances	Appliance power rating*	No. hours of use per week	AC or DC appliance?	Inverter losses**	Energy consumption per week
	n	P	t		L	(Wh)
-	-	-	-		-	n x P x t x L
Lights	4	12	4	DC	1	192
Hoover	1	500	0.5	AC	1.2	250
-	-	-	-	-	-	-
-	-	-	-	-	-	-
-	-	-	-	-	Total	442
-	-	-	-	-	-	-
*This should ideally be checked with a meter as discussed					-	-
**Inverter losses of 20% include battery charger losses, these don't apply to DC appliances						
This table calculates energy per week, but monthly or yearly figures could also be used						

Fig. 42. Calculating loads table
Courtesy of Rob Gwillim.

Device	Supply	Power
Halogen light	DC, 12/24V	5–20W
Fluorescent light	DC, 12/24V AC	8–20W 11W–40W
Fridge 225 litre 100 litre	AC AC	150–300W; 2–5kWh/day 100–200W; 1–2kWh/day
Fridge (100 litre)	DC, 12V, 24V	30–60W; 0.2–0.6kWh/day
TV, colour, new	AC	40–120W
TV, colour	DC, 12V	60W
TV, B&W	DC, 12V	10–20W
Music system	AC	30–50W
Portable stereo	DC, 6–12V	5–20W
VCR	AC	10–40W
Computer, desktop	AC	100–250W
Computer, laptop/ notebook	DC, 9–18V	10–30W
Printer, inkjet	DC/AC	15–50W, 2–6W standby
Printer, laser	AC	500–900W
Fax machine	AC	40–80W, 2–6W standby
Photocopier	AC	50W (idle), 700–1500W (operating)
Sewing machine	AC	50–80W
Blender, mixer	AC	100–300W
Vacuum cleaner	AC	500–1500W
Iron	AC	500–1500W
Washing machine	AC	300–800W (cold) 1500–3000W (hot)
VHF/UHV radio	DC, 12/24V	1–10W (standby) 25–80W (transmit)

NB. Most appliances carry a label or stamp detailing operating voltages and power consumption.

Fig. 43. Power consumption table.

Source: 'Off the Grid' by D. Kerridge & D. Hood, CAT Publications.

How to add up your load

1) Make separate lists of the AC and DC appliances you wish to use and consider which are essential and which are merely desirable (see priorities and flexibility on page 71).

2) Find the power rating of each piece of equipment (in watts), either from the maker's name plate on the back of the appliance, or by consulting the power consumption table here.

3) Estimate the average number of hours each appliance will be used each day for a rough guide, or use a meter for more accuracy (see below).

4) Calculate the number of watt-hours each appliance consumes by multiplying the power rating by the hours in use.

5) Factor in inverter inefficiency. An inverter loses a little electrical energy when it converts DC power to AC power so you will need to allow for these inefficiencies in your calculations. To do this multiply the AC watt-hours by around 1.2 (this may vary depending on the inverter you choose; check the manufacturer inverter specifications).

Go through these steps and you will come up with a total that gives you a fair idea of how much energy in watt-hours on average you will need to keep you going for a given period. In this example (see figure 42 on page 66) a renewable energy system is needed which will generate enough power to supply 442Wh of electricity each week.

Accuracy versus estimation

The power ratings labelled on many electrical appliances (for example, computers, televisions and so on) usually quote maximum demand rather than actual average usage, so metering of actual consumption is a good idea to assess energy needs accurately. Consumption can vary hugely between different people and according to lifestyles and habits, so using guesswork alone to predict how much energy will be used can be quite misleading.

Monitoring electricity consumption with a meter will often reveal that the appliances most costly to run are not the ones we expect. Sometimes groups of very small appliances that are left on 24:7 (like wireless internet routers and cordless phone charger bases) can use as much energy per week as a single large appliance, like a washing machine that is only used for a few hours a week.

Types of metering device

Three main types of metering device can be used:

1) Plug-in kWh meter. This has the advantage of allowing each appliance to be monitored separately but the disadvantage that only plug-in appliances can be metered, thus excluding fixed lighting, heating and cooker circuits.

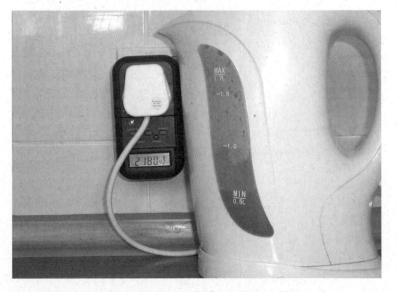

Fig. 44. Plug-in kWh energy meter showing a kettle using just over 2kW.

Fig. 45. Efergy wireless electricity monitor.
Image courtesy of Efergy.

2) Whole house electricity meter, such as the 'Efergy' wireless electricity monitor. This is a clamp-on sensor device which can be clipped onto the 'meter tails' between your electricity meter and consumer unit. The disadvantage of this type is that it meters the entire consumption of your house at once and this is a small inaccuracy in the readings because it assumes voltage rather than measuring it.

3) Smart meter. This is the type of meter now installed by utility companies in buildings with a mains utility supply. It is used to calculate electricity bills, but provides the user with more information. The UK Government plans to have smart metering in all UK homes by 2020.

TECHNICAL

Priorities and flexibility

To keep your system size down you can also choose to prioritise equipment according to critical need and optional use. When calculating expected consumption, use higher estimates for critical loads, and lower ones for optional loads.

One approach is to multiply estimated loads by a 'priority factor'. For example, 1.25 for critical loads, 1.0 for essential loads, and 0.75 for optional loads. The larger the factor, the greater the confidence you can have in continued supply.

Certain loads can be adjusted to suit the available energy – use lower estimates for these flexible loads.

Estimating the required size of your PV system from your load calculations

Having calculated how much energy your PV system needs to produce, a rule of thumb can be used to estimate (very approximately) what size PV system would be needed.

The irradiation can be found from PV-GIS or Meteonorm on the internet (see references). Note that the irradiation provided by PV-GIS or Meteonorm will be an average over many years and that it will vary considerably from one year to the next. An array should be generously sized to compensate for this.

This design process usually shows that stringent energy efficiency through careful architectural design, use of A⁺ rated appliances and other measures can make very good economic sense, as it reveals the saving from a reduced size solar PV array (see page 74 for section on energy efficiency in PV systems).

Batteries and stand-alone systems

The other key consideration when designing a stand-alone PV system is how much energy storage is needed in order to provide the required reliability (this is not an issue for a grid-connected PV system because the grid acts as a virtual battery – see Chapter Six). Detailed models which consider the length of the longest period with low irradiances subtracted by how long (if at all) the user can/will operate without electricity are used to assess how much battery storage is required. This gives the number of 'days of autonomy' on which a system must operate without sunlight.

Once you have worked out both the days of autonomy and average energy consumption figures you can calculate the size of your required battery store. This is something of an iterative process, since having calculated what size battery store is needed your model must then check what size PV array is required to keep it charged.

Calculating AC loads for peak demand

When calculating load you need to think about what would happen if everything were switched on at once – just as the national grid technicians have to make allowances for half time kettle surges during televised football matches. To do this, add up the power rating of each appliance to find the maximum number of watts. If there is any likelihood of this total switch-on scenario occurring (and actually it is quite unlikely), the inverter should be sized to cope.

Certain loads such as fridge compressors, motors and fluorescent lights require as much as three times their operating power for brief periods when starting up. Look for an inverter that can provide more than its rated output for short periods to cope with these start-up peaks.

A simple table to calculate maximum power demand is shown below. Note that in the case of larger systems, it can be assumed that not all the appliances and lights will be switched on at the same time.

Appliance description	Number of appliances	Appliance power rating*	-	AC or DC appliance?	Inverter losses**	Total Power consumption	-
	n	P	-		L	(Wh)	
	-	-	-		-	n x P x L	-
Lights	4	12	-	DC	1	48	-
Hoover	1	500	-	AC	1.2	600	-
-	-	-	-	-	-	-	-
-	-	-	-	-	-	-	-
-	-	-	-		Maximum Power consumption		
						648	Watts
System Voltage	12	Maximum Battery Current			-	54	Amps
		(Max Power/System Voltage)				-	

Fig. 46. Table: Calculating your maximum power demand
Courtesy of Rob Gwillim.

Using computer software to get a more accurate forecast

The above rules of thumb can be useful to make an initial assessment of the appropriateness and feasibility of a stand-alone PV system, however, there are many complex variables such as variations in cloud cover and shading from trees, which make calculating accurately very complicated. Thus to design a PV system which will perform satisfactorily, it is strongly advised to use one of the computer design packages, like PV-Syst or PV*SOL, or to use a professional installer. PVSYST is available from www.pvsyst.com; PV*SOL is available from www.valentin.de These design packages are expensive to buy, however, and require an understanding of PV systems engineering.

Energy efficiency and PV systems

Installing your own power system makes you more energy conscious. Suddenly you see your power supply in a whole new light. Because you are paying a huge amount for your electricity upfront – before you receive a single unit from your PV system – you have to maximise the benefits of your long-term investment by minimising waste. The less waste, the quicker the return on your investment. If you have old appliances in your house now might be the time to replace them with something more efficient. Or even to get rid of an appliance you can do without. This is especially the case with a stand-alone system where excessive demand might have to be compensated for by expensive and environmentally damaging diesel generators. Even if you plan to install a grid-connected system, where you have the flexibility to buy and sell from the national grid, it pays to work out your electricity demand in advance and find ways of making energy savings. Every watt saved is an extra watt sold.

The financial benefits of energy efficiency

Using the EU energy labelling system you can get a reasonable estimate of the total energy consumption cost of a domestic electrical appliance over its lifetime. The calculation is simple:

Annual consumption in kWh/yr x average cost of electricity £/kWh x estimated lifespan = total energy cost in £s.

Source: *Energy Saving House* by Salomon and Bedel, CAT Publications.

Take a refrigerator costing £150 and labelled with an annual consumption of approximately 240kWh/yr.
Estimated period of use: 15 years
Average peak electricity price per kWh electricity bought from the national grid: 0.14p

Cost of energy during period of use (assuming price of electricity doesn't change) becomes:

240kWh/yr x 15 x 0.14 = £504

This is three times more than the refrigerator costs to buy.

Now compare this to a second very efficient refrigerator using 120kWh/yr.

Cost in energy during period of use:

120kWh/yr x 15 x 0.14 = £252.00

The second fridge will save you £252.00.

See www.sust-it.net for a wide range of product energy efficiency comparisons based on the 2010 UK national average tariff of 13.97p per unit.

Maintenance

Of course, no renewable energy system will run without hiccups and disruption – although solar PV systems do require less ongoing maintenance than other renewable energy systems as there are no moving parts. To keep problems to a minimum you will have to maintain your system, and occasionally repair and replace parts, though damage to parts is likely to be infrequent. Some maintenance jobs are discussed in brief in later chapters of this book, additional maintenance requirements for stand-alone systems are covered in detail in the CAT Publication *Off the Grid*, available from www.cat.org.uk

Chapter Six

Grid-connected PV systems

Introduction

Under the regulations governing feed-in tariffs your PV installer will have to be registered with a Microgeneration Certification Scheme (MCS). There are a number of Microgeneration Certification Schemes run by national construction industry accreditation bodies. Your installer will be responsible for handling the relationship between you and the various bodies that manage the flow of electricity from your property to the grid. However, knowing how the national grid works helps you understand what PV installers are talking about when they mention terms such as public electricity suppliers and distribution network operator. This chapter is for people considering grid connection who want to know more about their obligations to others, safety concerns and technical requirements.

What is grid connection?

'Grid-connected' is a popular term rather than a technically accurate one. The national grid consists of overhead lines on large steel pylons moving electricity around at 400,000V and 600,000V; you definitely don't want to go anywhere near those with your PV system! When the term grid-connection is used in this book it is actually referring to the connection to the local distribution network. This is the local network which feeds off the national grid. When you supply electricity to the local distribution network from your PV system the electricity is used within that network.

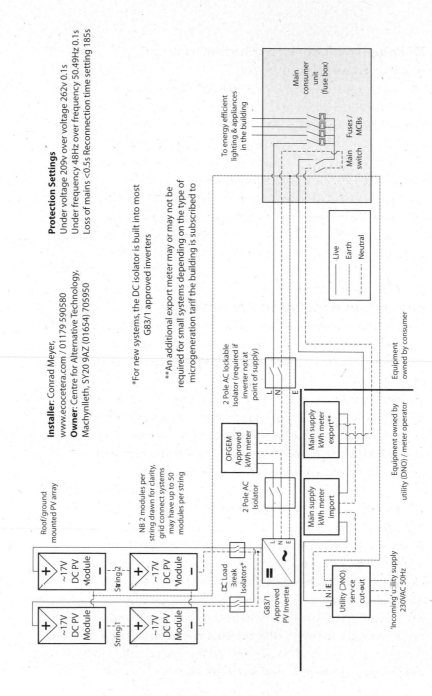

Protection Settings
Under voltage 209v over voltage 262v 0.1s
Under frequency 48Hz over frequency 50.49Hz 0.1s
Loss of mains <0.5s Reconnection time setting 185s

Installer: Conrad Meyer,
www.ecocetera.com / 01179 590580
Owner: Centre for Alternative Technology,
Machynlleth, SY20 9AZ (01654) 705950

*For new systems, the DC isolator is built into most
G83/1 approved inverters

**An additional export meter may or may not be
required for small systems depending on the type of
microgeneration tariff the building is subscribed to

Fig. 47. Grid-connected PV system schematic.

Differences between stand-alone and grid-connected systems

The local distribution network/'grid' is maintained by a distribution network operator (DNO), whose standards must be met before connecting your PV system. Once connected you will have a direct trading relationship with a 'public electricity supplier' (PES). The grid connection allows you the freedom to buy and sell electricity depending on your own needs and power production. There is no need for batteries, or load or charge regulators, unlike in a stand-alone system. However, you will need a different type of inverter (see page 83), and also Ofgem approved meters and other components.

Putting it all together – how a grid connected system looks

You can see from figure 47 that a grid-connected system has a similar overall layout to a stand-alone system but with some crucial differences. There are no load controllers and no DC appliances. The electricity passes straight though an approved PV inverter to be converted to AC ready for use in a normal 230V system and for exporting to the local distribution network. You'll also notice additional safety features such as DC and AC isolators. These are all specified under the regulations for grid-connected PV systems (see page 80).

The economics of grid-connected PV systems

When sizing a grid-connected PV system you can choose to match it to your aspirations. The size of a grid-connected PV system is not governed by the size of your load as with a stand-alone system. You can choose to generate a small or large proportion of your electricity needs depending on how much you want to invest. You can even choose to install a system greater than your needs and sell any extra electricity you produce. The feed-in tariff incentives (discussed in more detail in Chapter Nine) are likely to make this last option a more attractive proposition as it guarantees a fixed price for each unit of electricity produced over a long period.

Technical issues with grid-connected PV systems

Although financing is important, this chapter will outline the main technical and safety issues that have to be considered when a grid-connected system is designed and installed. First, it is helpful to understand the nature of the national grid network supply.

Embedded generation

Most of Britain's electricity infrastructure consists of a relatively small number of very large power stations powered by coal, oil, gas, nuclear and large-scale hydro energy. These feed power through the traditional network hierarchy to a very large number of small-scale consumers, as shown in figure 48.

However, renewable energy works in a slightly different way. Renewable electricity is usually produced by a large number of small generators, each of which is connected to (embedded in) the local distribution network. These electricity producers are called small-scale embedded generators (SSEGs). As soon as you install a PV system and connect it to your local distribution network you become a SSEG. This creates different safety issues to stand-alone systems, and is therefore subject to different regulations.

Rules governing grid connection

There are two sets of rules governing the connection of small-scale embedded generators: Engineering Recommendation G83/1 and Engineering Recommendation G59/1. These protocols are published by the Energy Networks Association, a body representing all DNOs. Systems above 3.68kW have

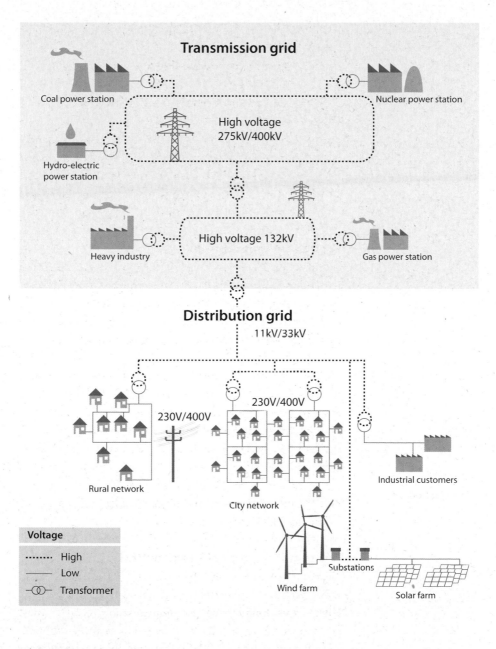

Fig. 48. Network hierarchy.

to comply with Engineering Recommendation G59/1. This requires a prior written application, system studies by the DNO and on-site witnessing of protection by the DNO's protection engineer. All these factors add considerably to the cost of a system. It is hard to justify these extra costs for systems only slightly larger than 3.68kW (or 11.04kW in the case of large commercial buildings with a 3-phase electricity supply).

For this reason, for most people it makes sense to have a generating capacity lower than 3.68kW. All systems with a generating capacity lower than 3.68kW are governed by Engineering Recommendation G83/1 (or G83/1 for short). Most people reading this book will not, and should not, attempt to connect to the grid without the help of an accredited solar PV installer, but knowing the regulations gives you a greater understanding of how your PV system works in relation to all other power sources operating through the grid.

The DNO has to be careful that all SSEGs connected to their part of the grid have 'Loss of Mains' protection that will automatically disconnect any PV system if there is a major electrical disturbance on the line. If there is a fault on the line and an embedded generator continues to supply electricity after the utility company has disconnected it at the substation (a situation called 'islanding') it is obviously dangerous, hence the need for automatic 'Loss of Mains' protection. In this kind of event it would be logistically difficult for a DNO to go round isolating every SSEG in the affected area.

The section below gives a summary of the G83/1 recommendation. Full details of G83/1 and G59/1 are available to download (for a fee) from http://2009.energynetworks.org/distributed-generation

Engineering recommendation G83/1

To summarise, the main constraints of G83/1 are:

- The inverter must be 'G83/1 type-tested'.
- The system must be no more than 16A (3.68kW) per phase (for an explanation of phase see single phase or 3-phase on page 86).
- There must be an isolating switch for the PV system near the point of connection which can be padlocked in the off-position only.
- The system must be labeled according to G83/1 for the safety of future maintenance personnel.
- The distribution network operator must be sent all relevant documentation once the system has been commissioned.

G83/1 inverter specifications

Under G83/1 you must use an inverter that has been type-tested by an approved laboratory; the G83/1 certificate for a PV inverter is usually published on the manufacturer's website. There are only a small number of G83/1 approved inverters on the UK market, due to the cost of getting each model tested. Hopefully in the future there will be greater harmonisation of PV inverter standards – with other countries having similar grid infrastructure – as this would mean a greater choice of inverters suitable for UK PV systems.

A G83/1 approved PV inverter will perform the following functions:

- Maximum Power Point Tracking (MPPT is usually a separate unit in a stand-alone PV system, but in a grid-connected system it is built into the inverter).
- Convert DC to AC.
- Export AC electricity with good power quality.
- Disconnect in the event of a power failure.
- Disconnect if voltage or frequency are out of acceptable limits.
- Display how much PV power is being generated.

Other features of some inverters, or optional extras:

• Display details of system parameters.
• Display weather conditions.
• Output data to home computer, building management system, or website.

Inverter sizing

The nominal rated power of a PV module is given at standard test conditions (STC). STC refers to an irradiance of $1000W/m^2$, cell temperature of 25°C and air mass of 1.5. Most of the time a module will produce less than this nominal power rating due to cloud cover and the sun being low in the sky. This is an important consideration when sizing an inverter and array for optimum performance. In the UK most of the solar energy we receive is at low light levels. This is shown as the solid line in figure 49.

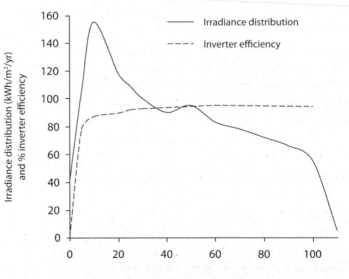

Fig. 49. Graph of UK irradiance distribution and inverter efficiency.

Fig. 50. OFGEM approved kWh meter with pulsed output.

Inverters, as with many energy conversion devices, operate at maximum efficiency when operating at full power, which means they operate inefficiently at low power (as shown as the broken line in figure 49). The key point here is that if we were to connect a 1kW inverter to a 1kW PV array, the inverter, would spend much of its time operating at a low efficiency and rarely at maximum efficiency. Hence – for the UK climate – it is usual to install PV inverters with a lower power rating than the array – by approximately 20%. PV inverters are designed for this, and conversely can operate above their rated power for short periods of time, though the equipment specifications should be checked carefully by the system designer.

Metering

The electricity meter connected to the existing incoming supply – the settlement meter – will not have been designed for import/export metering. Some meters will merely wind backwards if you connect your PV supply to the grid, which may be in breach of the contract with the electricity supply company. Some meters, including token meters, have anti-fraud protection and will permanently disconnect the supply if there is a reverse flow of electricity. Under FiTs your accredited PV installer will fit an additional meter next to the inverter that will meter the PV system energy produced. The main settlement meter will often also need to be changed. This has to be done by the meter operator under instruction from the electricity supply company.

Signing off a G83/1 connection and rules for multiple installations

A G83/1 connection form must be sent by your installer to the DNO once the system has been installed. If there is more than one system installed in a locality, your installer must make a prior application to the DNO, which will then carry out a system study to ensure the network is robust enough to cope with the altered usage; they may make a charge for this. Note that G83/1 is only a recommendation so minor local variations may apply according to the needs of the local DNO. An installer who has a lot of experience working in their local area should have a good relationship with the DNO.

TECHNICAL ————————————————————

Single phase or 3-phase

Electricity is the science of moving electrons around to make them work for us. The term 'phase' refers to the timing of the electron movements as they travel around, while a phase conductor refers to the part of an electrical circuit that is live. If wires are in phase, it means that timing of the electron movement back and forth, is the same. The electrons are 'in step' or 'in time'.

In general, most domestic dwellings, small offices or workshops have a single phase electricity supply, which consists of three wires: phase ('live'), neutral and earth. The phase and neutral wires provide a circuit to allow electrons to flow and energy to be delivered, the earth wire is used to bond all metal casings to prevent electric shock. A single phase system is used for these situations because relatively small amounts of energy are needed and it is cheaper to install.

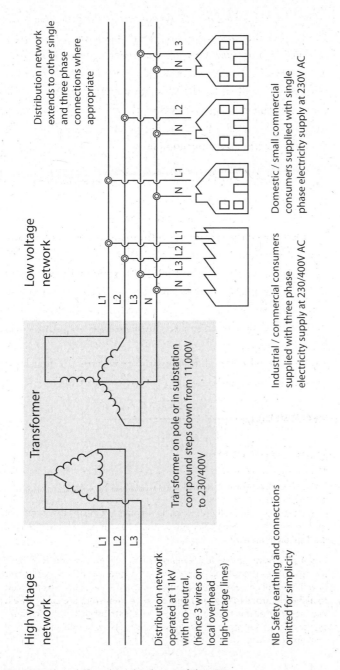

High voltage network

Transformer

Low voltage network

L1
L2
L3

L1
L2
L3
N

N L3 L2 L1

N L1

N L2

N L3

Distribution network operated at 11kV with no neutral, (hence 3 wires on local overhead high-voltage lines)

Transformer on pole or in substation compound steps down from 11,000V to 230/400V

Industrial / commercial consumers supplied with three phase electricity supply at 230/400V AC

Domestic / small commercial consumers supplied with single phase electricity supply at 230V AC

Distribution network extends to other single and three phase connections where appropriate

NB Safety earthing and connections omitted for simplicity

Fig. 51. Single and three phase distribution of electricity.

Industrial and large commercial properties usually have a 3-phase electricity supply. This consists of five wires: phase 1, phase 2, phase 3, neutral and earth. The three phase system is more efficient for delivering large amounts of electrical energy, but is more expensive to install.

Transmission and distribution of electricity is usually done using 3-phase.

At low voltage, both 3-phase and single phase connections can be supplied from the same circuit providing the power drawn from each phase is roughly equal (in other words, providing the phases are 'balanced'). It is because a DNO needs the phases to be balanced that we are limited to 16A per phase for an embedded generator, like a solar PV system. 16A (3.68kW) is deemed to be insignificant in terms of phase balancing. If you have a 3-phase connection, you can connect 16A (3.68kW) of inverter output to each phase (in other words 11.04kW in total).

When an inverter converts DC to AC it actually boosts the DC voltage to around 800V DC then electronically 'chops' it into AC. The section which boosts the voltage is called a boost converter. The boost converter works more efficiently if the voltage it receives (the input voltage) is closer to the voltage it sends out (the output voltage), because this way it has to do less 'boosting'. To make PV inverters work more efficiently, it is common to connect PV arrays with a voltage of 200-1000V DC. This also means that, for a given power rating, the cables will carry less current, so thinner cables and hence smaller switches, connectors, conduit, trunking and junction boxes can be used, all reducing the installation cost of a system.

SAFETY

Dangers associated with grid-connected PV systems

Grid-connected systems tend to have higher DC voltages than stand-alone systems. The disadvantage of this is that the voltages are high enough to cause a dangerous electric shock, unlike the 12V and 24V DC used in stand-alone systems. As such, the DC wiring must be carefully designed to prevent electric shock. This is done primarily by using Class II cables and connectors (double/reinforced insulation) or equivalent – you should see the double square symbol on any DC connectors and components used.

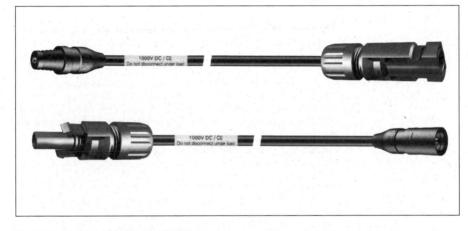

Fig. 52. MC4 (top) and MC3 (bottom) DC PV connectors commonly supplied with PV modules.
Image courtesy of Multicontact.

Fig. 53. Class II, double square symbol.
Image courtesy of IEC (International Electrotechnical Commission).

Differences between DC PV wiring and normal AC wiring

These are the three main differences btween DC PV wiring and normal AC wiring:

- Since PV modules always generate electricity in daylight, the DC wiring in a PV system cannot simply be switched off for maintenance or in the event of a fault, unlike normal domestic AC wiring.
- A DC isolator has to be fitted next to the inverter so that the electricity supply from the PV array can be disconnected.
- Any connectors and switches used in DC wiring must be specifically designed for DC use and for the voltage and current present. Normal AC connectors and switches MUST NOT be used.
- To prevent electric shock once the array has been connected, any testing or maintenance of the DC wiring between the array and the DC isolator must be done at night, or with the array covered with a secure thick, black opaque covering.

Fig. 54. DC isolator designed for PV systems; many new inverters now have a DC isolator built in.

• Because PV modules only generate a limited amount of current depending on irradiance levels, fuses or circuit breakers cannot be used to protect the cables from overheating and causing fires in the event of a fault. As such, the cables and other components have to be rated for the maximum short circuit current rather than the normal operating current.

This short summary is not an exhaustive list of precautions. For these consult *Photovoltaics in Buildings – Guide to the installation of PV systems* (2nd edition) published by the BERR (Department of Business, Environment and Regulatory Reform), formerly the DTI (Department of Trade and Industry) and available from the Energy Saving Trust website: www.energysavingtrust.org.uk

Fig. 55. AC isolator which is lockable in the off position only.

SAFETY

Isolating safety features

Isolation

There has to be a lockable AC isolator adjacent to the consumer unit so that an electrician can disconnect the PV system and padlock it in the off-position before carrying out any work elsewhere.

The PV inverter will also have a DC isolator either built-in or next to it. Both the AC and DC isolators are switched off whilst doing maintenance work on the inverter.

Safety labeling

Dual supply labels have to be fixed at the point of supply and to any other consumer units in the building.

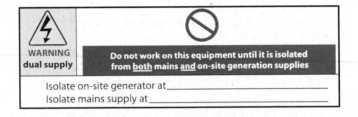

Fig. 56. Dual supply label.

Any DC junction boxes and isolators should be labeled 'Warning: contains live parts during daylight'.

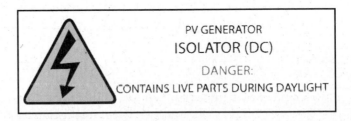

Fig. 57. A separate DC isolator label.

Chapter Seven

Array mounting systems

Introduction

By now you should know how many modules you will need to provide the power you want (array size), the array orientation (azimuth) and the angle you need to place it in relation to the horizontal plane (the tilt). The next step is deciding how it should be fixed to your roof, or an alternative mounting structure if the roof isn't suitable.

Roof mounted systems

There are three basic types of roof mounted structure:

- Integrated arrays – where the PV modules are used as an alternative to the roofing material (tiles or slates) on pitched roofs.
- Non-integrated arrays – where the PV modules are fixed on top of the roofing material on a pitched roof using one of several fixture designs.
- Flat roof arrays – where the PV modules are fixed to a flat roof using a framework or trough type system; this will almost certainly be angled to give the array the optimum tilt and orientation for maximum efficiency.

SAFETY ──────────────────────────────

Can your roof carry the load?

In the context of fitting a PV system to a roof the term 'load' refers to the amount of pressure exerted on the roof. There are two main loads that might be an issue with PV systems:

• Wind uplift loads – where the wind passing across the roof exerts an upward force on the modules, like a sail turned on its side.

• Dead load – the downward force applied on the roof by the weight of the modules and mounting system. In some cases the addition of a PV system can increase the amount of snow build-up on a roof, which also has to be considered.

If the wind uplift is too great it could cause the array to come off the roof, or – in high winds – even the roof to come off the building. The wind uplift can be reduced for pitched roofs by leaving a gap around the edge of the array, or by using a roof-integrated array. For older buildings, the roof has to be checked to make sure it is securely fixed to the walls of the building. More modern buildings generally have galvanised steel wall-tie bars that fix the roof trusses to the walls, but for older buildings these may have to be retro-fitted.

If the dead load is too great then it could cause crushing of the mounting system or roof structure. Dead load is more likely to be a problem with 'glass-glass' PV modules consisting of two layers of glass (in front of and behind the cells) as these are much heavier. However, the majority of modules have a layer of Tedlar™ plastic film at the back rather than glass, which is lighter.

Installation of small arrays onto modern domestic pitched roofs does not usually present any structural problems, but an experienced PV installer will be able to advise if there are likely to be any structural issues. Installation of large arrays onto commercial buildings invariably requires careful attention to structural issues. If there is any uncertainty over the suitability of the mounting required for a PV array, a structural engineer should be consulted. He or she will be primarily concerned with whether the roof and PV array structure are strong enough to withstand wind uplift. It is often preferable to ask the supplier of the preferred mounting system to recommend a structural engineer who is familiar with that mounting system. The structural engineer will carry out load calculations to local standards, and ensure that the structure can cope with the most extreme weather conditions likely in that area.

Pitch roofs

Rafters and battens

There is a lot of talk about rafters and battens in this chapter so if you're not familiar with roofing terminology you'll need to know the difference between the two.

• Rafters are the sturdy timbers (usually at a pitch of 30-60°) which form the main structure of any pitched roof. PV mounting systems must always be fixed to the rafters, not the battens. British Standard BS 8000-6: 1990 (part six) specifies the size of fasteners that should be used to ensure a secure fixing.

• Battens are smaller pieces of wood fixed across the rafters onto which a roofing material can be fixed (usually slates or tiles). On modern roofs there is always an additional layer of felt or plastic material called sarking which provides a second line of defence against dampness and draughts.

Flashing

Flashing is the term used to describe additional weather proofing at the interface between roof components. For example, where two different materials join at different angles, or at the base of a chimney where it meets the roof. In the context of a PV installation, flashing systems are occasionally used to seal around certain types of integrated PV mounting systems. However, the majority of mounting systems should not require flashing if installed correctly.

How temperature affects performance

In Chapter Three we focused largely on solar irradiance because it is the variable that affects system performance most. However, temperature also has a significant effect on performance. Many photovoltaic devices become less efficient as they get warmer and all of them become warmer when placed in sunlight. Because of this a PV array must be carefully designed to prevent the modules getting too warm. This is usually achieved passively using the natural property of air, which rises as it gets warm (convection); also, but to a lesser degree, by the wind.

In the case of an array that is bolted to the top of a roof (a non-integrated array) there is usually a generous gap behind and to the sides, which will keep it well cooled by these convection and wind currents.

Fig. 58. A non-integrated array powering lights in a public toilet, Beacon Hill, Loughborough.

However, where PV slates or tiles are used (a roof-integrated array) the air path behind and to the sides is restricted. The array will get hotter than an equivalent non-integrated array and produce less electricity.

Integrated arrays

Temperature control in integrated arrays The problem of module temperature can be improved in integrated arrays by 'counter-battening' (see figure 61 on page 98). This increases the air gap under the modules and allows for greater air flow. On an existing roof this would require all the slates or tiles on that pitch to be removed and re-fitted, a costly exercise unless a new roof is required anyway.

Even with counter-battening an integrated array will never perform as well as an equivalent non-integrated array. However, in many instances, for example conservation areas, national parks and listed buildings, integrated arrays may be the only way to get planning permission for a PV system.

Fig. 59. Installation of the flashing kit around a roof-integrated polycrystalline PV array with modules orientated 'landscape'.
Image courtesy of Dulas Ltd.

Fig. 60. Installation onto a slate roof of a non-integrated array on rails, with modules orientated 'portrait'.
Image courtesy of Dulas Ltd.

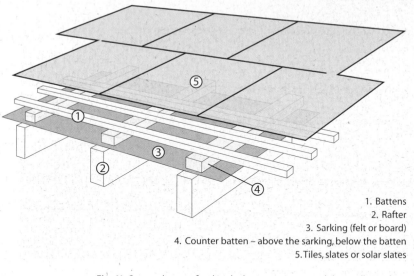

1. Battens
2. Rafter
3. Sarking (felt or board)
4. Counter batten – above the sarking, below the batten
5. Tiles, slates or solar slates

Fig. 61. Counter battens fixed under battens prior to module installation for a roof-integrated array.

Types of integrated PV modules

PV slates and PV tiles PV slates and PV tiles are available in different sizes. They are chosen to match the type and size of slate or tile already fitted to a roof. Despite the large number of modules needed, the use of factory fitted DC connectors means electrical connection is both quick and reliable. A number of PV tiles designed to be integrated with British roofing tiles are distributed by both Solarcentury and Lafarge/Redland. Each PV tile replaces several roofing tiles.

Some PV tiles are available with both polycrystalline and monocrystalline cells, according to aesthetic requirements.

PV laminates PV laminates are rigid modules supplied without an integral aluminium frame. A number of systems are available to mount laminates, usually consisting of an aluminium frame that is screwed down to the rafters. The frame has aluminium strips that clamp down onto the laminates with foam tape. A flashing material is provided to provide a weatherproof seal between the modules and the

Fig. 62. Installation of PV slates onto a counter-battened roof.
Image courtesy of solarcentury.com

Fig. 63. Solar Century C21 polycrystalline PV tiles installed on a heritage building.
Image courtesy of solarcentury.com

roofing tiles. However, the range of unframed laminates is declining as the PV market standardises towards aluminium framed modules.

Non-integrated arrays

SAFETY ————————————————————————————

- Careful design is required when installing PV systems on pitched roofs to avoid damage caused by wind uplift, which can cause the loss of PV modules and damage to roof structures if ignored. To prevent wind uplift there should be a gap between a non-integrated array and the edge of the roof.

- Fixings into rafters must be strong enough to withstand the loads but should not be greater than 10% of the rafter width. If you have 2x4 rafters (2" x 4"/50mm x 100mm) do not use a fixing with a diameter of more than 5mm. In some cases additional brackets or metalwork may be required. This is described in more detail in British Standard BS 8000-6: 1990 (part six).

Fixing a non-integrated array to the roof For existing roofs, non-integrated arrays are generally the cheapest for two main reasons.

- Standard modules can be used, which are competitively priced and quicker to install.
- Only a small number of tiles are removed – to permit fixing of roof hooks or stand-offs to the rafters (see below). The choice of roof hooks or stand-offs will be made by the PV installer according to what type of tiles or slates are on the roof. Generally roof hooks work best with tiled roofs and stand-offs with slate roofs, but there are exceptions.

In a non-integrated array system a series of connecting rails are usually fitted to roof hooks or stand-offs and the modules fitted to the rails. Standard PV modules have their own integrated rail running around the edge of the module (see figure 65 on page 102).

SAFETY

Roof hooks and stand-offs must always be fitted through to the rafters (not to the battens).

Roof hooks A roof hook is a bracket which is screwed onto a rafter. Fitting requires the removal of the roof tiles around the fixture point on the rafter. Once the roof hook is fixed the tiles can be replaced. One or more tiles will need to have a small notch cut into the back to allow them to sit correctly when placed back on the roof. The PV rail is then bolted to the exposed part of the roof hook which protrudes between two tiles (see figures 64 and 65).

Stand-offs Stand-offs are also bolted to the rafters. Again the tiles must be removed and replaced once the fitting is made. Each stand-off has a square plate with a protruding round tube. This tube is easily sealed using the associated flashing kit, which is about the size of the area of one tile. Welded to the top of the round tube is a flat plate onto which the array rails are bolted (see figures 66 and 67).

Fig. 64. Roof hooks installed on a tiled roof prior to fixing of rails.
Image courtesy of Dulas Ltd.

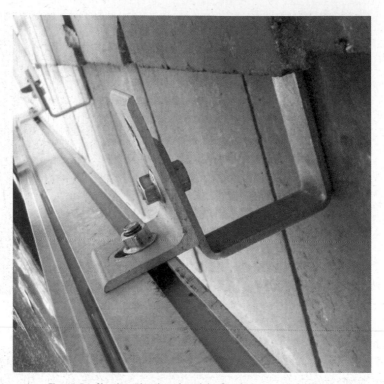

Fig. 65. Roof hooks with rails and modules fitted.

Fixing modules to rails Modules can be mounted either 'landscape' or 'portrait' (see figures 59 and 60 on page 97), wherever necessary in order to fit as many modules as possible within a limited roof space, but usually clamps must be fitted onto the long side of the module to prevent vibration and module fatigue (see figure 68 on page 104).

Non-integrated systems usually have some sort of rail (usually an aluminium extrusion). Where modules are installed in portrait position rails are fitted onto the roof hooks or stand-offs and run horizontally across the roof. The modules are then clamped onto the rails (see figure 68).

Where modules are mounted landscape the rails are fitted vertically. This will generally require more (albeit shorter) rails

Fig. 66. Stand-off and flashing kit during installation on a slate roof.

Fig. 67. Stand-off and flashing kits fitted to a pan-tile roof with Alutec rails and ready for module installation.
Image courtesy of Dulas Ltd.

(and hence increase the cost). This also means that the gaps between the modules are at the top and bottom rather than the sides making the collection of detritus like leaves, snow and even birds nests more likely.

SAFETY ——————————————————————

Installation of modules and mounting systems against manufacturers' guidelines may invalidate their respective warranties.

Fig. 68. Kyocera KC130 modules installed non-integrated on a Welsh Cedar shingle roof using UniRac rails.

Fixing thin film modules As discussed in Chapter Three thin film modules incorporate a thin film of semiconductor deposited on the glass front, rather than wafers of semiconductor. In these modules the solar cells are usually arranged, in strips (rather than in round, square or eight-sided cells, as with crystalline cells). These should ideally be orientated so the cell strips are vertical (which may be portrait or landscape according to type). This means any dirt build-up along the bottom edge of the module causes the same amount of shading on each cell thereby preventing module efficiency losses due to mismatches in the cell outputs.

Flat roofs

Flat roofs are most common on commercial office buildings, house extensions and garages. There are a number of systems available for fitting modules onto flat roofs.

Traditional frame techniques

The traditional method of mounting a PV array to a flat roof is to use a substantial aluminium or galvanised steel frame. This is either fixed to the roof structure – with careful attention to sealing – or weighed down with the addition of considerable ballast. This type of system allows the optimum tilt to be used, and generous ventilation for maximum performance. However as modules are angled up from the roof and so can behave something like an aeroplane wing exposed to the wind, the frame must be designed to withstand considerable wind uplift. The mounting must be designed so that the modules or frame cannot detach from the roof and the roof structure itself is not unduly loaded (subject to pressure which might cause structural damage to the building). In areas that experience heavy snowfall, there is a risk that the installation of PV modules could cause very localised snow-drifts beneath the modules, which would add to the weight on the roof. This can be mitigated by installing vertical screens at the back of

Fig. 69. Arrays fitted to a flat roof using Solkraft profiles.
Image courtesy of norsksolkraft.com.

each row of modules, or by leaving a gap between the modules and the roof.

Single module low profile mounting

The single module low profile mounting is an alternative to the heavy weighted frame and is generally faster to install.

A number of single module low profile options are available. These include:

- The 'ConSole'. This is simply placed on the roof and filled with concrete blocks or sandbags. The module is then fixed onto it. An advantage is that the module has a close to optimal angle of tilt and several can be laid out in rows all facing due south.
- The 'Solion'. Many flat roofs will not be strong enough to carry ballast weighted systems like the ConSole. An alternative is the Solion system. Solion uses an aerodynamic profile that uses wind from any direction to push the module onto the roof. The disadvantage of the Solion is a small reduction in performance due to the shallow angle of

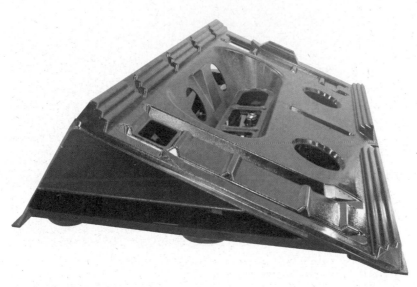

Fig. 70. Low profile mounting system for flat roofs.

Fig. 71. Installation of Kyocera KC130 Modules on a flat roof using the Solion system.
Image courtesy of Dulas Ltd.

tilt. Also, as a result, the modules will not self-clean (remove rain and snow) as well as modules on a steeper tilt.

Non-roof mounted systems

There are two main scenarios where it may not be possible to mount a solar PV array on the roof of a building.

1) Where planning permission is declined on aesthetic grounds, or because a house is listed or situated within a national park or conservation area.
2) Where concerns about roof strength show that the increase in weight or wind uplift due to a PV array could not be withstood.

Façades

A number of products exist to enable PV modules to be used vertically on the walls of buildings, for example to replace traditional materials in 'curtain wall' architectural façades.

Fig. 72. A polycrystalline PV façade at the Croydon Central shopping centre, with Proven wind turbine in background.
Image courtesy of solarcentury.com.

Fig. 73. Glass-glass laminates on pole mounts at the headquarters of inverter manufacturer SMA.

Where the PV modules can be competitively priced against the traditional façade materials, the cost saving often outweighs the disadvantage of a non-ideal angle of tilt.

Pole mounted and tracking systems

If it isn't possible to mount a PV array on a building, several ground mounting options exist. These often have the advantage of permitting optimum tilt, orientation and passive cooling; though a disadvantage is the potential increased shading from trees and buildings. Pole mounted systems may avoid planning approval problems by virtue of not being fitted to a building. They do, however, require robust foundations using wet concrete or screw anchors.

Tracking systems offer similar advantages and disadvantages to pole mounted systems, but with improved performance thanks to their ability to follow the sun's trajectory. However, tracking systems should be very carefully chosen for their robustness, as failed motors and controls could incur significant maintenance costs. Also, some types do not track very well during patchy cloud.

Using flat roof techniques on the ground

Ground mounted arrays can also be installed using the same systems as those used on flat roofs. The increased stability means the arrays can use simple ballast weights instead of foundations. However, to mount an array this close to the ground without chronic shading requires a large open space (and careful attention to lawn mowing as demonstrated in figure 74).

Fig. 74. Polycrystalline array on a rack at Beacon Energy, Loughborough.

Chapter Eight

Approvals – planning permission and building regulations

Introduction

Most small-scale solar PV systems installed on existing buildings do not require planning permission (though you should check with your local planning officer for the rules as they apply to your particular site, especially if you live in a national park or a listed building).

In general it is important to check which approvals are required from official bodies early on in the design of a PV system so that any necessary design changes can be made. Applications are usually made by the designer/installer of the system who will have a detailed knowledge of the regulatory requirements and how to comply.

The following is a checklist of approvals typically needed for a solar PV system. There may also be other local requirements.

Before installation

☑ Planning permission.

☑ Listed building consent.

☑ Building control (a structural engineer may be required to prove your building will not become a danger as a result of any changes made).

☑ Electricity supply contract (if you are connected to the grid).

☑ Distribution network operator (if you are connecting to the grid with a system larger than 16A per phase, or incorporating more than one property).

Fig. 75. Sunslate detail around a traditional dormer window.
Image courtesy of solarcentury.com.

After installation

☑ Inspection and testing of installation by a qualified electrician (see below under Building Regulations).
☑ Distribution network operator (if you are connecting to the grid).

Planning permission

Since planning permission can be the most contentious issue for renewable energy installations – possibly requiring negotiation – it should be a high priority when designing a system. The main issue that planning officers and committees are concerned with in the context of a PV array is aesthetics.

FiTs will make solar PV more familiar and understandable to local planning authorities but systems still have to comply with Permitted Development rules. These rules vary throughout Britain. In England and Scotland the installation of solar PV is covered by Permitted Development rights introduced on 6th April 2008 in England and 12th March 2009 in Scotland. In Wales and Northern Ireland Permitted Development is a devolved responsibility and both governments are currently considering changes to their legislation to facilitate installations of microgeneration technologies. Until then people living in Wales and Northern Ireland must consult their local authorities. Your PV installer will quickly be able to identify whether planning permission will be a problem for you but its useful to see what the issues are for yourself.

The governments planning portal website provides lots of useful information:

- http://www.planningportal.gov.uk/wps/protal/ genpubLocalInformation for a database search of planning authorities.
- http://www.planningportal.gov.uk/england/public/ planning/ for more general information.
- For full details of the 2008 planning amendments visit http://www.opsi.gov.uk/si/si2008/pdf/uksi_20082362_ en.pdf
- For a very good summary of the different planning requirements in England and Scotland visit: http://www.energysavingtrust.org.uk/Generate-your-own- energy/Getting-planning-permission

If you are unclear about whether your proposed solar PV array constitutes permitted development or if it would require planning permission, it is advisable to check with your local planning officer. Send them a brief note describing the system and a sketch showing the proposed layout on the roof. Alternatively, contact a planning consultant, architect or PV installer with expertise in this area.

If your property falls within one of the following, then you would definitely need to apply for planning permission/listed building consent, and convince the planning officer of the aesthetic merits of your system:

1) In a conservation area and your property is visible from a public right of way.
2) In a designated area: national park; Norfolk Broads; Area of Outstanding Natural Beauty (AONB); or Site of Special Scientific Interest (SSSI).
3) Is within the curtilage of a listed building.

These permissions/consents do not necessarily prohibit PV systems, but the onus is on you to convince the planners that your system will blend in with its surroundings or have exemplary architectural merit. Popular strategies are either to use PV tiles or PV slates, which are similar in appearance to the existing roof. Another strategy is to try a free standing array (see Chapter Seven for advantages and disadvantages).

Historically, planning decisions regarding renewable energy have been somewhat inconsistent and have tended to focus

Fig. 76. Use of C21 Solar Tiles to blend in with a heritage building, Clutterbuck's Funeral Home.
Image courtesy of solarcentury.com.

heavily on aesthetic considerations at the expense of wider regional and national commitments to reduce carbon dioxide emissions under the Kyoto Protocol. The UK government has set out national renewable energy policy in a planning context and standardised some of the issues by publishing *Planning Policy Statement 22 (PPS22)* on renewable energy, as well as a companion guide with more technical detail.

PPS 22 states that local authorities should give encouragement through positively expressed policies regarding small-scale renewable energy developments, and that these policies should include targets for renewable energy capacity in the region set for 2010 and 2020. However, if a target has been met, this should not in itself be used as a reason for refusing planning permission.

Whilst this is only a guideline, it does make a useful document to refer to in applications and appeals. Where local authorities are in breach of the guidelines, this should be reported to central government as feedback on compliance with PPS22. Local authorities, such as Merton, Croydon, Sheffield and Reading, now require 10% of energy on major developments to be delivered by renewables.

Merton Council has a policy incorporating the following text: 'All new non-residential development above a threshold of 1000m^2 will be expected to incorporate renewable energy production equipment to provide at least 10% of predicted energy requirements' (see 'Merton Rule' references and further reading).

If planning permission or listed building consent is required then good communication and building up a good relationship between all parties can go a long way in achieving a result which is satisfactory to all concerned.

Building Regulations

The following aspects of a PV system require Building Regulations approval:

1) Spread of fire to adjacent properties – Part B.
2) Wind uplift and structural strength – Part A & BS 6399 Part Three (check with a structural engineer if unsure).
3) Electrical wiring installation – Part P.
4) Accessibility (for example to controls) – Part M.

Any electrical work on a dwelling house or property incorporating a dwelling must now be done by an electrician registered with one of the following: BRE/ECA, BSI, CORGI, ELECSA, NAPIT, NICEIC, or else be notified to the local building control department. This includes work on both AC and DC circuits in a PV system, regardless of whether it is stand-alone or grid-connected.

In general, building control departments in the UK tend to be overstretched, so often the onus will rest on the designer and installer of a PV system to ensure compliance in order to limit personal liability incurred from any future problems.

DNO approval

For a single installation of less than 16A generating capacity, which is not part of a wider development, you should not normally need to make a prior application to your DNO, providing the inverter is G83/1 approved. G83/1 type approval certificates are usually published on the inverter manufacturer's website. The installer must, however, submit comprehensive G83/1 commissioning documentation (also see Chapter Six).

Chapter Nine

Financing a solar electricity system and feed-in tariffs

Whilst a solar electricity system will save money on future electricity bills, there is a lengthy gap between the capital investment in the system components and installation and whatever financial returns are made during the life of the system.

So just how much money will you have to pay up-front? And how will you pay for it?

Costing a PV system

There is no fixed price per kWh you can expect to pay for the installation of a solar PV system. All the factors discussed so far will change the cost – from tilt and orientation to shading and efficiency losses. The competitiveness of installer and PV module costs will also be a factor, as will the type of roof access you have and your choice of roof fixings. You could create a budget by gathering component costs as described throughout the book. However – bearing in mind that the cost of the system will define whether you go ahead or not – the best and most precise solution is to ask several installers to give you a quote (see Chapter Ten).

A simple breakdown of costs based on an actual stand-alone PV system

To give you an idea of the costs involved, here is a simple cost breakdown of a 1.25kWp grid-connected system:

- Solar PV modules – 10 x Kyocera KC125 (Watt) modules
  = 1250Wp: £5,000
- Mounting System: £900
- PV system electrical installation: £1000
- Inverter: SMA Sunny Boy SB1100 (1100W): £1022
- PV system total cost: £7,992

Does solar PV make economic sense?

In broad terms solar PV, or any other form of renewable energy, makes economic sense when an electricity source is needed and the costs of the alternatives are greater; for example, the cost of connecting to the national grid or buying and operating a diesel generator. If you are already connected to the national grid and your goal is to cut your CO_2 emissions by consuming only green electricity, it is cheaper in the short term to buy your electricity from a green electricity supplier.

In the future the cost per watt of PV system installation will fall as the manufacturing and installation processes become more streamlined and the volume of components manufactured increases. Conversely, the cost of electricity bought from the grid – and hence the value of electricity generated by a solar PV system – is likely to rise, as the price of fossil fuels rise, especially in the long term once short term fluctuations are averaged out.

However, at the moment some form of subsidy is required to level the cost of solar electricity with that from other sources.

Subsidies for PV systems

Until recently, encouragement of renewable energy by the UK government focused on the so called Renewables Obligation - which obliges public electricity suppliers (PESs) to buy a percentage of energy from renewable sources using a complicated trading scheme based around the fluctuating price of Renewable Obligation Certificates (ROCs).

In the past there were also capital grants for installation but these have all now been replaced by feed-in tariffs (see below).

The advantage of a capital grant was that it used to help offset the additional cost of interest on commercial loans or loss of interest on any savings used. However, the advantage of feed-in tariffs is that it provides solar PV system owners with a guaranteed price for each unit of electricity they produce over many years – making it easier to guarantee any loan.

ROC's continues, but now for only the largest systems. Most people reading this book will not come into contact with them. If you are planning a larger system and are thinking about claiming ROCs it is useful to know that the main disadvantage with them is that the subsidy price can fluctuate depending on market conditions (see ROCs section on page 124).

Feed-in tariffs

Feed-in tariffs started in Britain in April 2010. Under the scheme every single unit of electricity generated by solar PV over a 25 year period attracts a payment of up to 41.3p.

As this amount stays the same no matter how much electricity prices fluctuate, it provides small scale renewable energy generators with a guarantee that the costs of installation will be recovered, even when taking out a loan. They have a regular income from their system, as well as the savings they are making from generating their own electricity. Because there is a direct reward for every unit of electricity produced feed-in tariffs also encourage renewable energy system owners to keep their system operating at their optimum with close attention to design and maintenance.

How do they work?

Despite the name, the new FiT is actually a generation tariff. Generators will be paid for every unit of renewable electricity they produce, whether they use it themselves or sell it back to their electricity supply company. Even stand-alone systems, will be eligible for FiT income. As our example below illustrates there are three sources of income or savings under the new scheme:

- Generation tariff income for all electricity produced - up to 41.3p per kWh for PV (<4kW).
- Export tariff income from electricity exported to the grid - 3p per kWh
- Electricity bill savings from using own electricity instead of buying electricity from the grid; typical savings could amount to 14p per kWh if the electricity can be used directly at the time when it is produced (though its worth pointing out that prices rise and fall).

Who is eligible? The scheme is open to all major forms of renewable electricity generation (solar PV, wind, hydro, biomass and anaerobic digestion), up to 5MW (5,000kW) capacity, so long as renewable energy system owners use Microgeneration Certification Scheme (MCS) eligible products installed by MCS accredited installers.

What about stand-alone PV systems? Even if you're not connected to mains electricity you can receive income from the generation tariff (in addition to savings in diesel) – as long as you certify that all electricity generated has been used.

What about existing PV systems? Eligible generation installed before July 2009 will only receive a generation tariff of 9p/kWh – which is about the same income they would have received under microgeneration tariffs available prior to FiTs.

Will FiTs make a difference? The FiT scheme only applies to small-scale generators – those with less than 5MW installed nameplate capacity. Realistically, the bulk of renewable energy in the UK will be produced by large-scale, industrial generators such as wind farms which the government supports in different ways. But while FiTs may not drastically change the UK energy mix, they will make a real difference for individuals and communities who want to take low carbon transition into their own hands.

As the following example calculation shows, FiT income makes small-scale renewable energy start to look very good, not only in ecological but also in financial terms.

At this level it becomes incredibly important that a PV system generates the electricity you hope it will, so choosing an installer who will hand-over a system that delivers on its promise is vital (see Chapter Ten).

FiT example calculation: The photovoltaic family

Estimating PV yields Naturally, the economic viability of a PV system is very much dependent on how much energy it produces. A solar installer will give you a figure predicting how much electricity your system will produce annually. If they get this figure wrong and the actual output is less than the predicted output it will result in a direct loss of income. Verify that the installer has predicted system output using a reliable design package, such as PVSYST or PVSOL, and that they will install the system to a high standard using quality components to ensure its reliability. You can use the PVGIS online tool (http://re.jrc.ec.europa.eu/pvgis/apps3/pvest.php) for approximate estimates of PV yields.

In the calculation below, we assume an annual production of 800 kWh per 1 kW of rated PV capacity. This is a typical value for a south-facing, unshaded roof in central England.

The solar family In 2010, a family installs a solar PV array with a maximum output of 2.5 kW on their south-facing roof. The system costs £14,500. This includes a 25 year warranty extension for the inverter unit (costs for inverter warranties/replacements should not be overlooked in financial calculations for domestic PV). The roof produces around 2,000kWh of electricity per year. The family use 50% of the electricity directly at the time when it is produced and sell the other 50% on to the grid.

As a retrofit system – in other words one installed on an existing roof – with less than 4 kW capacity, the generation tariff for their roof will be 41.3p per kWh. This rate will be guaranteed for 25 years, with index-linked changes to reflect inflation.

The annual income from their PV system can be calculated as follows:

Generation tariff: 2,000 kWh x £0.413/kWh = £826
Savings on electricity: 1,000 kWh x £0.12/kWh = £120
Export tariff: 1,000 kWh x £0.03 = £30
Total annual income and savings = £976
The system will pay back its capital cost in a little less than 15 years, and from 2025 to 2035, the family will receive almost £1,000 (at 2010 values) per year as tax free annual profit.

This calculation does not factor in the loss of efficiency over time as PV modules age. Manufacturers of quality PV modules usually guarantee that their products will still provide 80% of their rated power after 20 (or 25) years. The calculation also doesn't factor in any increases in domestic electricity prices. If electricity prices rise (and it seems very likely that they will) then the savings from generating your own electricity increase and the pay-back time shortens.

How high are the tariffs? Tariff levels are banded by technology and by scale (generation capacity). The table below shows a selection of tariffs. For example, electricity from a domestic PV roof (below 4kW) will receive a subsidy of 41.3p/kWh (unit) of electricity if retrofitted to an existing property, whereas the same amount of electricity from a 500kW wind turbine will receive 18.8p/kWh. Export tariffs will be the same for all: 3p/kWh.

Generation tariff payments will be index-linked (in other words they will increase with inflation) and are guaranteed for 25 years for PV (20 years for wind and hydro).

Annual degression To provide an incentive for reducing the cost of renewable technologies, the government will lower the generation tariff for newly installed generators in future years. This is called 'annual degression'. For example, a home owner who installs a retrofit PV roof in 2012 will only receive a generation tariff of 37.8p per kWh instead of the 41.3p they would have been eligible for if they had installed the system in 2010. However, this will not affect a home owner who installed a retrofit PV roof in 2010. They will still receive the 2010 rate for the full 25 years. Once a project is registered, the only change to the tariff will be an annual increase index-linked to inflation.

Technology	Tariff (p/kWh)
PV (<4 kW) (new build)	36.1
PV (<4 kW) (retrofit)	41.3
PV (4-10kW)	36.1
PV (10-100kW)	31.4
PV (100kW-5MW)	29.3
PV free-standing	29.3
Hydro (<15 kW)	19.9
Hydro (15-100kW)	17.8
Hydro (100kW-2MW)	11.0
Hydro (2MW-5MW)	4.5
Wind (<1.5kW)	34.5
Wind (1.5-15kW)	26.7
Wind (15-100kW)	24.1
Wind (100-500kW)	18.8
Wind (500kW-1.5MW)	9.4
Wind (1.5MW-5MW)	4.5

Fig. 77. Feed-in tariffs for different renewable energy systems installed in 2010 and 2011.

Claiming FiTs All eligible systems have to be registered in a central database maintained by the energy regulator Ofgem (or sub-contracted out to a specialist provider). Before registration a PV system must meet the eligibility criteria and be accredited as follows:

- For systems under 50kW the installer must be registered with the Microgeneration Certification Scheme (www.microgenerationcertification.org).
- For larger systems accreditation uses the same process as the Renewables Obligation.

Tariff payments are co-ordinated with regular payments of electricity bills. The electricity supplier pays the feed-in tariffs (and bills users for the electricity imported in the normal way). Suppliers may choose to 'net these amounts off' and just pay a cheque or submit a bill for the difference. Claimants can also appoint an agent to collect the tariffs on their behalf. Generally, the payments should be made quarterly to coincide with electricity bills. Your installer will help you claim FiTs.

As stand-alone generators do not have a direct relationship with an electricity supplier they will have the right to approach any large electricity supplier (mandatory FiTs supplier), who will be required to pay their FiTs.

For more information about FiTs visit CAT's website www.cat.org.uk/fit or www.fitariffs.co.uk

Renewable Obligation Certificates (ROCs)

Under the new FiT rules, generators under 50kW can only claim the feed-in tariff, not ROCs, so this will apply to the majority of UK PV systems. Generators between 50kW and 5MW will have to make a one-off choice between FiTs and ROCs. Those generators who installed their PV systems before July 2009 have to be registered for ROCs in order to claim FiTs.

Public electricity suppliers (PESs) are obliged to procure a percentage of their electricity sales from renewable sources, while the renewable electricity generator is eligible to claim a Renewable Obligation Certificate (ROC) for every megawatt-hour generated. The electricity supply companies must demonstrate that they have bought the required amount of renewable electricity by purchasing certificates for it. A PES could supply all their customers with power generated at their own power stations running on fossil fuels, but doing so would mean having to buy ROCs from another generator.

ROCs are traded on an open market, so like any commodity their value can rise and fall according to supply and demand. Demand is increased when the government increases the target percentage of renewable electricity sold to consumers. Supply is increased as more renewable energy generators are installed.

Increase in the value of property

A study by the US National Renewable Energy Laboratory (NREL) found that zero energy homes attracted a higher purchase price than other comparable homes because the value of the PV system and any savings or income derived from it stay with the property. If a house is required to have a home information pack (HIP) including an energy performance certificate (EPC), the actual and potential energy rating of the house will be shown. These factors will make micro-generation a bigger selling point in the future. There are also proposals for a reduction in Stamp Duty on the sale of 'zero carbon' homes, though the definition of a zero carbon home is likely to be contentious; the use of zero carbon in this context often focuses on the carbon emissions associated with building use and ignores the emissions generated in its construction or demolition.

Cost of electricity

Clearly the cost of electricity has risen significantly in recent years as the UK's oil and gas reserves are now in decline. Oil and gas will increasingly be imported at higher prices. A recent report by the Energy Watch Institute in Germany found that world oil production peaked in 2007. As production declines and demand grows prices will keep rising. However, energy markets are very volatile, so whilst one can be reasonably confident of the increasing value of home electricity generation for personal decision making, it would be hard to borrow money for a PV system based on a financial plan that relied on rising electricity prices.

Energy savings

The financial benefits of energy efficiency are described in detail on pages 74-75 in Chapter Four – Stand-alone systems. If you haven't already read this section it is worth going over because energy saving measures take on a new significance when you own a PV system, especially if you are claiming FiTs. Every unit of electricity saved is potentially a unit sold back to the grid.

Renewable electricity supply contracts

Under feed-in tariffs all the larger public electricity suppliers will be obliged to participate and hence offer the FiT rates to customers with grid connected renewable energy systems. Your PV installer will take care of these arrangements but you will have to sign a contract with a PES.

Chapter Ten

Finding an installer

Introduction

It is beyond the scope of this book to describe the entire design and installation process of a PV system in full as both stand-alone and grid-connected PV systems have complex safety issues requiring specific design and installation. It assumes that in the vast majority of cases readers will call on the services of a professional PV installer. So what does a PV installer do?

What does a PV installer do?

A PV installer will be your main point of contact, and see your project through from feasibility study to commissioning (switch on) with you. They will work out whether your site is suitable for PV, how many PV modules you will need, what the predicted output will be, how much the whole system will cost; and then, if you give them the go ahead to complete, they will buy the components and install the complete system. If you are connecting to the grid they will also handle the relationship between you and the various people you will need to deal with at the public electricity supplier (PES) and distribution network operator (DNO), already described in Chapter Six. They will also complete any paperwork required for G83/1 and FiT approval. Your installer should provide a warranty on materials *and* labour, and be available for maintenance in the future. Of course, businesses come and go, so it is preferable to find an installer who has been on the scene for a while and looks set to continue. Every year you will want to check that the actual output of your PV system matches the predicted output the installer gave you after the first feasibility study (see Chapter Twelve).

What should I ask potential installers?

Everything in this book so far should prepare you to ask your installer the right questions, but most people will probably ask the following three types of question first, before getting down to the nitty gritty of solar irradiance and tilt angle:

- How much will my PV system cost and how long will it take for it to pay for itself? Your installer will be able to give you a complete financial breakdown of costs and benefits, including FiT returns.
- How much maintenance will it need? How much will I have to pay for that maintenance? How much of it can I do myself? Installers should be able to describe average call-out charges as well as annual running or replacement part costs.
- How much good will I be doing for the environment? Most people are installing solar because they want to make a positive contribution to the environment. Installers should be able to advise you as to how much CO_2 you will save over the lifetime of your PV system.

Of course, there will be lots of other things you will want to ask. Just remember, you are the client making the significant investment. You have a right to ask as many awkward questions as you like. Generally speaking installers are a reliable, accredited bunch keen to engage with clued-up customers. They will also want to see that you start off with a good knowledge of your system so you make the most of it and tell your friends and neighbours how well solar works for you.

How to identify the best installers

Obviously, when spending a large sum of money on a project like this it's important to have confidence in your installer. You need to know that the money you invest will be spent wisely on a system that is well designed and installed by competent people using quality equipment.

The feed-in tariff (FiT) requires all renewable energy installers to be registered with a UK micro-generation installer certification scheme; there are a number of schemes listed at www.microgenerationcertification.org PV installers have to satisfy rigorous criteria in order to register, including:

☑ Appropriate health & safety policy(ies).
☑ Be registered with the 'REAL' quality assurance scheme.
☑ Be a competent electrician.
☑ Use calibrated equipment where appropriate.
☑ Install systems in compliance with the DTI's
   PV in Buildings – a guide to the installation of PV systems.
☑ Prove competency in PV installation – for example, by
   passing the City and Guilds 2372 Course: 'Installing and
   Testing Domestic Photovoltaic Systems'.

The competencies required by the BRE for solar PV installation under their micro-generation certification scheme are as follows:

• Structural engineering (understanding of static and wind
  uplift loads and impact on building structure).
• Roof work (understanding of roof construction and
  integrity issues such as rain penetration).
• Working at height.
• Solar resource assessment.
• Non-conventional electrical properties (for example,
  variable voltage according to module / ambient
  temperatures).
• DC electrical systems.

• Conventional AC electrical systems (BS7671).
• Grid connection requirements (G83/1, G59/1).

PV installation presents many new challenges regarding DC wiring, embedded generation and working on roofs that electricians won't usually be familiar with. There is now a City and Guilds course for electricians (mentioned above): the

Fig. 78. Electricians learning how to install a PV mounting system at the Centre for Alternative Technology.

'Certificate in Installing and Testing Domestic Photovoltaic Systems' (2372), see www.cat.org.uk/courses and www.cityand guilds.com

The former British Photovoltaic Association is now part of the Renewable Energy Association (www.r-e-a.net). The site has a list of PV installers that are REA members and are bound to the REA code of practice. The installer should be approved under Building Regulations Part P (see Chapter Eight), if the system is to be installed on a dwelling or building comprising a dwelling.

If you want a specific brand of equipment installed you could contact the UK distributor for that brand. The main distributors often have a network of installers around the UK who have often received specific training in the use of their products.

Having contacted an installer, you could ask about other systems they have installed in your area. Ask if you can make contact with the owner (to see if they are happy with the service they've received) and go and have a look at the system in situ.

Checking your installer contract

The following is a useful check list when the time comes to sign a contract:
- ☑ Installer can be found on www.microgenerationcertification.org
- ☑ Installer can provide evidence of Part P approval to do electrical work on dwellings.
- ☑ Installer has visited the site in person to carry out a survey for design, including an analysis of shading, roof tilt and orientation, and type of electricity supply arrangement.
- ☑ Quote includes detailed list of all equipment to be installed.
- ☑ Quote includes datasheets, manuals and warranty details (not just brochures) for main components (modules, inverters, mounting system).
- ☑ Equipment can be found on www.microgenerationcertification.org

☑ Quote includes a detailed layout drawing of the proposed array.

☑ Quote includes a detailed prediction of system performance based on industry standard design packages

☑ Contract places responsibility on the installer to get approvals for Building Regulations, planning permission, G83/1 (DNO approval) and feed-in tariff as appropriate.

☑ Contract places responsibility on installer for all electrical, mechanical and structural design and installation work.

☑ Contract includes warranty on materials AND workmanship.

☑ Contract specifies that installer will replace any defective equipment at THEIR cost.

☑ Quote includes comprehensive handover pack on CD or hardcopy as you wish, including all drawings, calculations, approvals certificates, datasheets and manuals.

☑ System includes an easy means for you to check instantaneous power and energy generated to date.

☑ Quote is fully inclusive of all design, approvals, materials, labour, making good, tidying up, travel and subsistence, commissioning, handover and warranty, with no hidden extras.

☑ Installer can give you addresses for a number of reference systems they have installed nearby – not just photos on a website.

Ideally

☑ Contract includes performance guarantee over the life of the system (for example, 20 or 25 years).

☑ Contract specifies terms and rates (index linked to inflation) for ongoing maintenance.

☑ Contract allows for you to retain, for example, 10% of contract cost until 30 days after commissioning so you have leverage to resolve any initial problems.

What happens if things go wrong?

Installers should have a complaints procedure in order to gain MCS accreditation. If you are dissatisfied, ask to see it. If after making a complaint you are still dissatisfied, notify the body operating the MCS scheme with which your installer is registered and contact a solicitor, law centre or citizens advice bureau for advice on how to proceed.

DIY installation?

There is no such thing as DIY installation of solar PV systems. This is not the sort of thing you can buy in a flat-pack from Ikea and assemble on a Saturday afternoon. There are health and safety hazards with the potential to cause serious injury or death, from falling off roofs to electric shock. The wiring in PV systems is more dangerous than conventional domestic wiring because of the high DC voltages involved and because the array wiring will always be live during daylight. There will inherently be parts of the array wiring that cannot be switched off except by working at night or by covering the array with a robust and completely opaque black sheet. Hence, if you are seriously considering installing your own solar PV system, you need to achieve equivalent competencies to those required by the BRE above. This would basically mean becoming a qualified electrician and doing a PV installers course, so is unlikely to be justifiable unless you're planning to do a lot of electrical work.

Chapter Eleven

Safety, installation and handover

Introduction

This chapter does not intend to provide an exhaustive list of safety precautions required for a PV installation, but rather an introduction to the key issues.

The main safety issues with PV installation are:

- Working at height.
- Electric shock.

Working at height

Working at height is governed by the Working at Height Regulations 2005. The main points are as follows:

Duty holders (in other words you, installers or anyone who has a responsibility for ensuring the health and safety of people connected to, or coming into contact with your project) must:

- 'Avoid work at height where they can' – for example, by cleaning a PV array on the roof of a bungalow with a long brush instead of climbing a ladder with a cloth and bucket.
- 'Use work equipment or other measures to prevent falls where they cannot avoid working at height' – for example, by using scaffolding to install a PV array.
- 'Where they cannot eliminate the risk of a fall, use work equipment or other measures to minimise the distance and consequences of a fall should one occur' – for example, by using harnesses where protection by scaffolding is not possible. See www.hse.gov.uk for further information.

Pitched roofs

Cherry pickers (hydraulic lifts used by firemen, builders and others to work safely at heights) are not usually suitable for use in PV installation because the installer can only reach a short distance from the caged working platform, and the working platform can only be moved slowly around the roof.

By contrast a pair of roofers or PV installers can move around a roof which has perimeter scaffold protection very efficiently. Hence, for most sloping roofs, a FULL scaffolding system will be required to install a PV system.

Note the word full. Even for the smallest 6 square metre (3m wide by 2m high) 600Wp array – in the UK the optimum array size for the smallest inverter available is 600Wp – a scaffold tower alone would not be wide enough to protect the installers from falling off the roof.

Scaffolding is a significant part of the installation cost, so it is worth trying to combine a PV installation with roof, chimney or other building repairs.

Fig. 79. A typical scaffold system used for PV installation training at CAT with handrails, walkboards, toeboards, brick guards, netting and access ladder.

Fig. 80. Temporary handrail for safe working on a flat roof.

Flat roofs

In the case of an installation on a flat roof, things are much easier. Temporary handrail systems which use weights as ballast might be suitable. This would be much cheaper than a full scaffold system.

Responsibility

You will need to be clear with the PV installer about which of you will be responsible for organising scaffolding. Scaffolding is always erected by a recognised and qualified scaffolding company, not by the builder or installer. When the scaffolder has finished the install they will fit a 'scaffold tag' to verify it is safe to use. This confirms their details, signature and installation date.

In addition to protection from falls, the scaffold contractor and PV installer will also be responsible for protecting other people nearby, by preventing falling objects and so on.

Electric shock

The electric shock risk from a PV system presents new challenges to an electrician because the cables from the array will always be live during daylight hours (since the sun cannot be switched off). The voltage from most crystalline modules is usually within 'Extra-Low Voltage' (ELV) limits (50V AC or 120V DC). However, once the modules are connected in series the voltage will be 200-1000V DC for a small grid-connected PV system. This is higher than normal domestic wiring.

All connection boxes and switches in a PV system should normally have special labels to warn that the connections are always live during daylight.

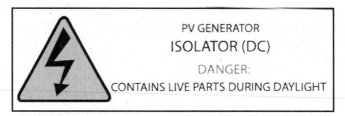

PV GENERATOR
ISOLATOR (DC)
DANGER:
CONTAINS LIVE PARTS DURING DAYLIGHT

Fig. 81. Warning label on DC isolator.

Specific procedures and equipment are available to install PV systems safely, and specialist training is also required (see Chapter Ten).

Since the PV installer will be a qualified electrician and will often need to connect the PV system to the main consumer unit (fuse board), it might be logical to have any other minor electrical work done at the same time.

TECHNICAL

PVC and environmental safety

The production, use and disposal of PVC results in the creation and release of persistent toxic and carcinogenic chemicals, including dioxin – a hormone disrupter and one of the most toxic substances known. By using non-PVC cables and components – for example, halogen-free or rubber insulated cables in place of PVC – this environmental contamination can be avoided. Talk to your installer about using the best environmental options.

Commissioning

Instantaneous performance check

The plane of array

The 'plane of array' is the angle at which the slant of a roof points towards the sun. As solar PV modules do not have to be fixed to a roof the plane of array can refer to the angle of slant on a separate structure built specifically to house solar PV modules – in which case you can build the optimum slant into the design of your structure.

Measuring plane of array irradiance

To measure plane of array irradiance you'll need an irradiance meter and a place to use it. These cost upwards of £200 new so most people will prefer to leave it to their PV installer to check. Have a look at figure 82; the sensor part of the unit is aligned with the plane of the roof where a PV array will be mounted (in this case an outbuilding on a rural farm). The reading is only 575W/m² because it is 4pm on a September afternoon, and the roof faces due south – the sun is now off to the side of the roof (the plane of array). If the irradiance meter was pointing directly at the sun, the reading would be

Fig. 82. Measuring solar irradiance with a hand held irradiance meter.

much higher. Measure at noon to get the best reading for your site, and then at different times of the day to remind you your module will not always be receiving peak irradiance. Once you have measured the plane of array irradiance you can make a rough estimate of what the array power output should be.

Your irradiance measurement will usually be somewhere between 0 and 1200 W/m², although exact values will depend on cloud cover, shading, time of day and time of year.

Approximate values you might expect to see in the British Isles are shown below:

0-50 W/m² – Night-time.

50-100 W/m² – Very overcast sky at beginning or end of day.

100-300W/m² – Cloudy/overcast sky.

500-800W/m² – Clear sky, sun low in sky.

800 -1000W/m² – Clear sky, sun high in sky.

1000-1200W/m² – Exceptionally clear day, very bright sunshine in summer.

As discussed in Chapter Two, module and array power output are rated at standard test conditions, which are $1000W/m^2$; 25 degrees Celsius and Air Mass 1.5.

So the expected array power can be estimated using the following calculation:

Expected array power (P_{EST}) $\simeq$

Rated Peak Array Power (PMPP) x $\dfrac{\text{Measured Irradiance } (G_{PLAN})}{\text{Irradiance at STC } (1000W/m^2)}$

For example if we have a 1.5kW PV array and an irradiance of $450W/m^2$ is measured, then:

Expected array power (P_{EST}) $\simeq 1.5kW \times \dfrac{450W/m^2}{1000\ W/m^2}$

$$\simeq 0.675kW$$

$$\text{Or } 675W$$

($\simeq$ stands for 'approximately equal to')

Note that this is a very rough approximation, values will increase or decrease according to module temperature and solar spectrum. Below about $200W/m^2$ it is very difficult to predict what the power should be with any accuracy, even with expensive scientific instruments. Also remember to allow for 5-10% losses when the inverter converts from DC to AC. Most inverters display AC output power not array DC power output.

A PV installer should usually carry out a more detailed version of the above measurement and calculation, whereby the irradiance and module temperature are measured, and the voltage and current produced (Voc & Isc) by the array are adjusted for actual conditions. These are then compared with the module datasheet, to identify any faults in the modules or wiring. The PV installer should include a summary of these test results in the handover pack.

Handover

The installer should provide the following documents at PV system handover. These documents should be kept in a robust folder located by the inverter so they can be referred to by anyone working on the PV system or other building electrical systems in the future:

☑ Installer contact details.
☑ System schematic including AC and DC parts.
☑ System operating instructions including procedures to shut down or re-start the system.
☑ Electrical installation certificate (BS7671 17th Edition Certificate with additional PV amendments), and the following attachments: Inspection Schedule and Schedule of Test Results.
☑ Copy of G83/1 SSEG Installation Commissioning Confirmation.
☑ Inverter Datasheet, Manual, G83/1 Test Certificate and Warranty details.
☑ PV Module Datasheet, Installation Manual and Warranty.

Additional optional items include:
☑ Structural engineering calculations, if appropriate.
☑ Details of any Building Regulations, planning permission or listed building consent received for the system.
☑ Details of any grants applied for.
☑ Details of the feed-in tariff used to sell electricity generated by the PV system, including any requirements for meter readings or other submissions to the electricity company.
☑ Details of any monitoring system and instructions showing how to access the data.

Chapter Twelve

Performance monitoring

Some sort of monitoring is essential in any PV system, in order to check periodically whether the system is producing the output it should be, and so that in the unlikely event of a fault it can be rectified promptly and without significant loss of generation. Monitoring can range from a simple bar graph on the inverter through to a comprehensive monitoring system with public display or system website (for example www.sunnyportal.com).

Monitoring equipment

Ofgem approved kWh meter
All UK systems should be fitted with an Ofgem approved kWh meter. This is similar to the meter you'll have as your main electricity meter. It gives a calibrated and non-volatile display of kWh and is very resilient to power failures, surges and so forth. It is usually fitted near the inverter, and gives a reading of solar energy generated.

Inverter displays
The inverter itself will have some sort of display, from a simple bar graph, to complex menu system, depending on the model.

Most inverter manufacturers offer a comprehensive monitoring system as an optional extra:

• Weather sensors can be added to assess whether the system performance matches climatic conditions.
• Connection can be made to a home or office PC or laptop to download comprehensive data and display it in a variety of formats.

Fig. 83. Display on a Fronius IG40 inverter at CAT.

- PV systems can be connected to the internet by telephone line or broadband, so that it can be viewed on the internet on a webpage operated by the inverter manufacturer; this can be set as public or private depending on your wishes!

Public displays

A public display is sometimes required, for example for systems in schools and public buildings. The signal for the display comes via a simple two wire connection to a dedicated pulse output from the PV system's kWh meter mentioned above. These displays usually have options to display three or four readings. For example:

- Present power output (W).
- Energy produced to date (kWh).
- CO_2 reduction since install (kg/CO_2).
- Value of electricity generated (£).

For kg of CO_2 the ratio is programmed according to the national figures for average carbon emissions from electricity production (for example, 0.43kg of CO_2 per kWh for the UK – Carbon Trust).

Because of the way these displays integrate and average the

Fig. 84. A public display in a school showing: present power output (kW), energy generated to date (kWh) and CO_2 emissions saved (kg).
Image courtesy of solarcentury.com

pulse signal, they are not always very accurate and the figures on the public display may not be identical to the display on the inverter. However, in most cases the display units are housed in different locations so this isn't a problem.

Scientifically accurate data

If accurate data is required for scientific research purposes, a custom-built monitoring system will be required, with accurate sensors measuring system and environmental variables including:

- AC/DC voltage.
- Current.
- Irradiance.
- Ambient and module temperature.

The sensors will be connected to a reliable, calibrated datalogger which will record values at, for example, 5 second, 10 second or 1 minute intervals; the data is then downloaded – manually or automatically – to computer software, which is programmed to calculate power, efficiency, performance ratio and so on.

Fig. 85. EETS monocrystalline reference cell and pt100 temperature sensor in radiation shield on a 13.5kW monocrystalline array at CAT.

Checking performance

A number of different performance indicators are used to compare the performance of different PV systems. The term efficiency is often heard in the context of PV, but often when efficiency figures are quoted for solar PV systems they just refer to the cell efficiency and don't compare how well the rest of the system has been designed – for example, how much energy is lost in the cables or how much is lost due to shading of the arrays by trees and other buildings.

A common ratio used to compare overall system performance is kWh/kW/year. A typical figure for a well designed system on a south facing roof at 35° pitch with little shading might be: 750kWh/kW/yr. But an accurate predicted performance ratio figure should be provided by the installer.

This ratio is very useful to the PV system owner. The predicted output given when the system was quoted for can easily be checked by taking readings once a year from the PV system's kWh meter (so make sure you get one fitted), then dividing the value by the size of the system. This figure is also useful because, from the predicted figure you are given when you buy the system, you can easily calculate how much you will save on your electricity bill each year, depending on which tariff you are on.

A worked example for an energy yield calculation

A homeowner installed a 2.5kW system three years ago and wants to check how well the system is performing and whether it is performing as predicted in the quotation provided by the installer.

The installer cautioned that the system would be slightly shaded by a tree in an adjoining garden but only during the winter months, and that the performance would be slightly reduced because the roof is facing south-west. In the initial quotation the installer predicted the system would generate approximately 710kWh/kW/year.

The homeowner wrote down the initial reading from the Ofgem approved meter (located next to the inverter) when the system was installed. They have now re-read the meter three years later, on the anniversary of system installation. These are the figures recorded:

• Meter reading on 21st June 2005 = 000006
• Meter reading on 21st June 2008 = 005069

The homeowner then carries out a number of simple calculations:

• Electrical energy generated by system   = meter reading 2 - meter reading 1
     = 005069 - 000006
     = 5063kWh (units)

• Electrical energy generated per year   = energy generated ÷ no. years
     = 5063kWh ÷ 3 Years
     = 1687.7kWh/yr

• Total yield   = Electrical energy generated per year ÷ system nominal power rating
     = 1687.7 ÷ 2.5kW
     = 675kWh/kW/yr

This system is performing below the 710kWh/kW/yr predicted by the installer.

However, the period monitored had abnormally wet summers. Over a longer period annual variations in performance would average out, and the average output since installation would be closer to any value predicted by a computer model based on weather data collected over many years. In general, the system is performing reasonably well considering the non-optimal roof orientation, shading and weather. There is no cause for concern.

Chapter Thirteen

Operation and maintenance

Grid-connected systems

A key advantage of grid-connected PV systems is that they require very little operation and maintenance, as all the day to day functions of the inverter are automatic. However, it can be interesting and rewarding to have a look at what the system is generating periodically. It is a good idea to check the system once a week, so that any faults are quickly solved and no energy is wasted. Some inverters can be setup with monitoring systems that can be checked over the web and can even send you automated emails if there is a fault.

Stand-alone systems

Stand-alone PV systems will almost certainly require more careful operation to ensure electricity is available when required. The battery voltage must be checked often (for example, twice a day) and some types of battery require the electrolyte (sulphuric acid) to be periodically checked and topped up with distilled water. This requires careful adherence to health and safety requirements for battery stores since the sulphuric acid is highly corrosive and the batteries can release explosive hydrogen gas during charging. The details of stand-alone system control – in other words, operation and maintenance – is outside the scope of this book, but useful guidance can be found in *Off the Grid* by D. Kerridge & D. Hood available from CAT Publications (www.cat.org.uk).

Concerns common to both grid-connected and stand-alone systems

Wiring

Since a PV system falls under Building Regulations Part P and BS7671 (IET wiring regulations), it should be periodically inspected by a qualified electrician, as with any fixed electrical wiring (see Chapter Nine). How often this inspection should take place depends on the type of building and its usage. For example, residential accommodation must be inspected every ten years or at change of use; for commercial property it is required every five years. For public buildings, construction sites and so on shorter periods apply: see the IEE BS7671 17th edition wiring regulations for further details. The periodic inspection and testing will include detailed examination of all electrical fittings and testing of all cables.

Array cleaning

Generally, arrays on pitched roofs will be steep enough to self-clean, particularly by hail and snow during the winter. The thin film of dirt that will build up only reduces output by a few per cent and access onto roofs will usually be too expensive to justify cleaning. The ideal tilt for PV systems in the UK is 30°-40° (see Chapter Three). In some circumstances it may be more economic to make a compromise on the pitch and mount an array at a much flatter angle, but due to the increased need for cleaning this should not be considered where the array would be difficult or expensive to access safely.

Any array which can be safely accessed without specialist access equipment should ideally be cleaned annually to prevent build-up of dirt and moss, particularly along the frame at the base of the module. If cleaning requires a cherry picker or scaffold tower, the improvement in electricity production due to the reduction in minor shading is unlikely to justify the cost of annual cleaning, so the array should be cleaned at no more than five yearly intervals. When cleaning a PV array, great care should be taken not to scratch the glass. The use of high pressure jets of water, which might penetrate into electrical connections, should also be avoided. Cleaning should be done with a bucket of warm water with mild soap (for example washing-up liquid) and a soft cloth.

SAFETY

The high voltages present in a PV array could be a hazard when cleaning if there is a fault in the wiring or the modules – for example, if cables have been damaged by rodents or building work – and there could be a risk from electric shock from the array frame. If there is any doubt as to the condition or history of a PV array, it would be advisable to have the array inspected and tested by a competent PV installer before cleaning is carried out.

Annual routine check

An annual routine check of a PV system is advisable so that any issues that might compromise system performance or require warranty claims can be dealt with promptly. Ideally the checks should be made by an experienced PV system installer who will be aware of what to look out for and be able to make a competent judgement based on the need for remedial action. This annual check could also incorporate cleaning.

The annual checks required will vary depending on type of system, but might include:

The array

☑ Check array hasn't become excessively shaded by trees, shrubs, or other obstructions: is there a large shadow over the array for an hour or more each day?

☑ Visually check for nests (birds, insects or rodents), which might shade the array or inhibit ventilation, and remove if appropriate. For arrays on domestic pitched roofs this can often be done from the ground; for flat roofs access to the roof will be needed.

☑ Check for any cracks, discoloration or other degradation of PV laminates or modules that might require replacement under warranty.

Inverter

☑ Check inverter location isn't excessively hot or dusty (or damp).

☑ Check for dust build-up on or around inverter (switch off inverter and remove dust with vacuum cleaner).

☑ Check inverter is adequately ventilated (has no obstructions) as per inverter manufacturer installation manual.

☑ If inverter has a fan (which will run only on sunny days) does it run smoothly or make noises, suggesting bearing failure?

☑ Record the reading on the electricity (kWh) meter next to inverter and calculate annual production (see Chapter Eleven). Compare values with predictions made by PV system installer.

☑ Check values shown on inverter or monitoring system are as expected from the performance predicted by the installer.

General

☑ Check that handover documentation is present (usually kept by the inverter).

☑ Check wiring and fittings for damage/discoloration.

☑ Check signs and labels are secure and legible.

☑ Check AC and DC switches to disconnect the inverter are accessible and clearly labeled as to their function.

☑ Stand-alone PV systems require additional checks: battery condition, electrolyte level (depending on type of battery). Also check correct operation of any manual and automatic controls.

☑ Depending on tariff, check necessary readings have been sent to your public electricity supplier (PES), and any payments for PV electricity sold have been received (for tariff details see Chapter Nine).

☑ Other checks as recommended in wiring regulations.

Calling out an installer

As mentioned at the beginning of this chapter, a PV system should be checked by the owner at least once a week (or daily for systems over 50kW) to verify that it is generating as expected for the weather conditions. If there is a fault, or the performance is consistently poor even on sunny days, the installer should be contacted to discuss the problem, with a view to a possible site visit to check the problem.

Chapter Fourteen

Case studies

These case studies are presented to provide an overview of the typical components of PV installations and their costs. Both domestic and commercial grid-connected systems are detailed in addition to a domestic stand-alone system. The focus on systems in Mid-Wales (the area in which CAT is based) was for ease of interviewing owners and getting photographs of the systems, and doesn't reflect the average distribution of PV systems in the British Isles, nor the type of building upon which they are fixed. It is far more common to find solar PV installations in urban areas, both on new build as well as a wide range of older properties.

Case study 1:
Stand-alone wind/PV/diesel system
Location: **North Wales**

Fig. 86. Pole mounted PV array and Proven WT2500 wind turbine.

Installer:	Dulas Ltd
Funding:	£10,000 from Energy Savings Trust PV Demonstration Programme Stream 2 (funded by the Government). Remainder by owner
Cost:	£28,000
Installation date:	2007
Modules:	8 x Kyocera KC125 modules (polycrystalline silicon)
Array:	1000Wp at STC[1], pole mounted
Array pitch:	45°
Array orientation:	0° (due south)[2]
Charge controller:	Studer Tarom
Wind turbine:	Proven WT2500 (2.5kWp)
Batteries:	24 x BAE 1340Ah (48V DC system)
Inverter:	Studer HP Compact 48VDC: 230VAC 6kW
Inverter undersizing:	Not relevant for stand-alone system
Monitoring:	Basic monitoring built into charge controller & inverter
Annual energy generated:	No record available[3]
Expected performance:	780kWh/kWp
Actual performance:	No record available[4]
Planning permission:	Required due to national park/listed building
Electricity tariff:	N/A (stand-alone/off-grid system)
Energy used for:	All electric lighting appliances in buildings, in addition to a borehole pump and heating controls. Energy also from diesel and wind power

Notes:

- [1] STC = Standard test conditions
  (irradiance = $1000W/m^2$, cell temperature = $25°C$, solar
  spectrum = AM1.5), see Chapter Two.

- [2] For azimuth measurements, south is taken as 0°, west
  as 90°, east as 270° as opposed to conventional compass
  bearings which take south as 180°, east as 90° and west
  as 270°. See Chapter Two.

- [3] Annual energy generated was checked by recording the
  reading on the electricity meter installed next to the
  inverter, and dividing it by the number of years that the
  system has been running for (for example 18 months =
  1.5 years). The method assumes that a new kWh meter
  was used which started at 000000 kWh, which should
  usually be the case.

- [4] Performance ratio kWh/kWp is simply the energy
  generated per year divided by the array peak power
  rating (kWp) at STC.

Note that the battery store & inverter have been included
in the PV system costs, because the aim is to demonstrate the
total cost of a stand-alone PV system. In practise the battery
store is shared with the wind turbine. A slightly larger battery
store, but the same sized inverter would have been required to
operate the wind turbine without the PV array or vice versa.

Breakdown of costs:		
Solar PV modules		£3,850
PV pole mount		£870
PV system miscellaneous components: includes Tarom charge controller		£500
PV system electrical installation		£640
Inverter (converts battery voltage to mains voltage)		£1,590
Batteries		£4,550
PV system total cost		£12,000
	VAT	£4,00
	System total	£28,000
Capital grant (£2,500 for PV & £2,500 for wind.)		-£5,000
Upfront cost to owner		£23,000

Background

A stand-alone system was installed because the property was too remote for an affordable grid connection. By installing the PV and wind systems the residents were able to make major savings on the fuel and running costs of the existing Lister diesel generator, and reduce noise. It also means they have modest amounts of power 24 hours a day 7 days a week, so they can now run a borehole pump and automated central heating. This would not have been viable with the diesel generator alone.

Planning permission was required for the PV array and wind turbine because the property lies within a national park. The property is a listed building, so listed building consent

Fig. 87. From left: 48V battery store, inverter, wind turbine charge controller.

was required. The owner was advised that consent wouldn't be granted for a PV array fixed to the building, so a pole mounted system was installed away from the house. The pole mount increased the cost of the system, but has the advantage that the array could be pointed due south at optimum tilt, away from shading objects and be well ventilated for optimum performance.

The array is tilted at 45 degrees to optimize winter output when the sun is lower in the sky but more energy is required for lighting. This tilt angle is at the expense of summer output, when less energy is needed. A battery can only store energy in the short term and not between seasons. If this system had been connected to the grid (where electricity can be bought and sold according to need) the array could have been tilted at 30-35 degrees for best all year round performance.

Case study 2:
Grid-connected commercial PV system

Location: Welsh Institute for Sustainable Education
(WISE), Centre for Alternative Technology,
Machynlleth, Mid-Wales

Fig. 88. 6kW PV array above a covered seating area using ROMAG glass-glass laminates.

Installer:	Dulas Ltd
Funding:	£42,000 from Energy Savings Trust PV Demonstration Programme Stream 2 (funded by the DTI). Remainder from other funders.
Cost:	£62,000 (excluding construction of roof structure)
Installation date:	2007
Modules:	78 x Romag 84.3Wp custom semi-transparent monocrystalline silicon
Array:	6.58kWp at STC, roof integrated
Roof pitch:	34°
Roof orientation:	12° (south by west)
Inverter:	1x Fronius IG30; 1 x SMA Sunny Boy; 1 x Mastervolt
Inverter undersizing[1]:	22%
Monitoring:	Ofgem kWh meter & scientific monitoring system
Annual energy generated:	3172kWh /Year
Expected performance:	750kWh/kWp/yr
Actual performance:	482.4kWh/kW/yr – calculated from kWh meter
Reasons for difference:	The roof has to be switched off for use by installer training courses
Planning permission:	Variation to main construction project
Electricity tariff:	£0.10/ kWh saving on commercial electricity bill + £0.03 /kWh ROCs (Large commercial site means most PV electricity is used on site). CAT will switch from claiming ROCs to FiTs from 2010 onwards
Energy used for:	All electric lighting, computers, appliances, heating controls in buildings with any surplus exported or shortfall imported from grid connection

Notes:
- [1] For PV systems in the UK the peak power of the array (kWp) is generally designed to be slightly larger than the inverter nominal AC power output P_{ACNOM} (kW). This is to allow for the fact that irradiance in the UK climate is on average less than the STC value of $1000W/m^2$ and because inverters are more efficient when they are worked harder.

Background

The Centre for Alternative Technology (CAT) is both the publisher of this book and a solar PV courses provider amongst many other things. It has just built the Wales Institute of Sustainable Education (WISE) to provide a world class educational centre for post-graduate students wishing to enter a career in an environmental industry. It has a history of generating renewable energy for its own consumption dating back thirty years but it now aims to be a net exporter of renewable electricity to the grid. The new WISE centre has created a growth in the demand for energy. Biomass CHP, solar thermal and solar PV systems were specified to compensate. The roof above a covered outdoor seating area was chosen for the PV array, because the roof faces close to due south and has a 34 degree pitch. It is close to the optimum tilt and orientation for a UK grid-connected PV system. Being a covered and unheated space meant that the PV array could be generously ventilated to keep the PV modules as cool as possible and operate them at optimum efficiency. Due to the size of the system, three separate inverters were used, as this improves performance. Normally when multiple inverters are used, they are all of an identical size and type, but because this array is used for teaching and research, three different brands of inverter were used to allow a comparison to be made. The application for planning permission was submitted as a variation to the planning permission for the main building project, which had already been granted.

Case study 3:
Grid-connected domestic PV system
Location: North Wales

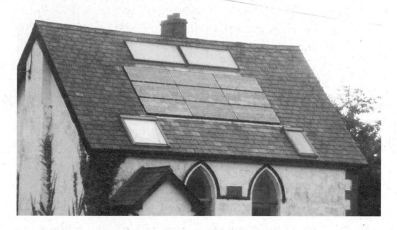

Fig. 90. 1kW Polycrystalline array on a slate roof in North Wales (the nine central panels) plus two solar thermal collectors above them.

Cost:	£12,143
Funding:	DTI/EST PV Grants
Installation date:	June 1999
Modules:	9 x Kyocera KC120 polycrystalline silicon
Array:	1080Wp at STC; Non-integrated
Roof pitch:	43°
Roof orientation:	-26° (south-south-east)
Inverter:	Fronius Sunrise Micro P_{NOM} =750W
Inverter undersizing:	22%
Monitoring:	Ofgem kWh meter & Fronius datalogger with PC connection
Expected performance:	760kWh/kWp/yr
Actual performance:	568kWh/kW/yr – calculated from kWh meter
Reasons for difference:	The system had been switched off for maintenance

Planning permission:	Required due to roof alteration not due to PV array
Electricity tariff:	£15/quarter estimated – Good Energy; Excludes saving on electricity bill
Energy used for:	All electric lighting, computer, appliances, heating controls in the home buildings with any surplus exported or shortfall imported from grid connection

Background

The owner of this small Welsh cottage was fortunate in having a roof which faced almost due south. They already had a solar water heating collector on the roof and chose to supplement this with a solar PV array. Planning permission wouldn't normally have been required, but other maintenance to the roof resulted in an overall increase in the roof height so the PV array was included in the application.

References and further reading

All web links are correct at time of going to press, but may change. If links no longer work try shortening them to the homepage address. For example: http://www.bsi-global.com/en/ProductServices/Electrical/Electrical-installers/ becomes http://www.bsi-global.com. Then use the site's search box to look for your document (for example product services, electrical installers). Alternatively, look for your document using a search engine like www.yahoo.com or www.scroogle.org, experimenting with different keywords if you don't see it the first time.

Books

CAT Publications

Kerridge, D. & Hood, D. (2008) *Off the Grid*, CAT Publications.

Helweg-Larsen, T. & Bull, J. (2007) *Zero Carbon Britain*, CAT Publications.

Piggott, H. (2006) *Choosing Windpower*, CAT Publications.

Salomon, T. & Bedel, S. (2007) *The Energy Saving House*, CAT Publications.
www.cat.org.uk/catpubs

Government (BERR) PV publications

BRE, SunDog, (2001) *Photovoltaics in Buildings: Guide to the installation of PV systems,*

DTI URN:02/788.

Halcrow, (1998) *Photovoltaics in Buildings: Testing, Commissioning and Monitoring Guide,* DTI URN: 98/1101.

Max Fordham & Partners, (1999) *Photovoltaics in Buildings: A Design Guide,* DTI URN: 99/1274.

Studio E Architects, (2000) *Photovoltaics in Buildings: BIPV Projects,* DTI URN:00/1516.

BSRIA, (2000) *Photovoltaics in Buildings: Safety and the CDM Regulations,* DTI URN:00/1517.

DTI, (2002) *Guidance on the Electricity Safety, Quality and Continuity Regulations.* DTI URN: 02/1544.

Econnect & ILEX, (1999) *Technical Guide for Connection of Embedded Generators to the Distribution Network,* DTI ETSU K/EL/00183/REP.

Studio E Architects, *Photovoltaics in Buildings: A survey of design tools,* DTI ETSU S/P2/00289/REP.

Terence O'Rourke, *Photovoltaics in Buildings: Town Planning Considerations,* DTI ETSU S/P2/00304/REP.

Department of Business, Environment & Regulatory Reform (formerly the DTI) publications from: www.dti.gov.uk/publications

Other publications

Antony, F.; Durschner, C.; Remmers, K. H., (2007) *Photovoltaics for Professionals: Solar Electric Systems Marketing, Design and Installation,* Earthscan Publications Ltd.

German Solar Energy Society; Ecofys, (2005) *Planning and Installing Photovoltaic Systems: A Guide for Installers, Architects and Engineers,* Earthscan Publications Ltd.

Green, M. A., (2000) *Power to the people, sunlight to electricity using solar cells*, Australia: University of New South Wales Press.

Hermannsdorfer, I. & Rub, C., (2005) *Photovoltaics: Old Buildings, Urban Space, Landscapes,* Jovis Verlag.

Komp, R. J., (2002) *Practical Photovoltaics: Electricity from Solar Cells,* USA: Aatec Publications.

Markvart, T.; Castaner, L., (2003) *Practical Handbook of Photovoltaics: Fundamentals and Applications,* Elsevier.

Markvart, T., (2000) *Solar Electricity (UNESCO Energy Engineering Learning Package),* John Wiley and Sons Ltd.

Messenger, R. A. & Ventre, J., (2003) *Photovoltaic Systems Engineering*, USA: CRC Press Inc.

Palz, W.; Greif, J., Heidelberg (1996) *European solar radiation atlas.*

Prasad, D.; Snow, M., (2005) *Designing with Solar Power: A Sourcebook for Building Integrated Photovoltaics,* Australia: Images Publishing Group Pty.Ltd.

Sick, F.; Erge, T., (1996) *Photovoltaics in Buildings: A Design Handbook for Architects and Engineers,* James & James (Science Publishers).

Solar Energy International, (2004) *Photovoltaics Design & Installation Manual, Renewable Energy Education for a Sustainable future,* USA: New Society Publishers.

Strong, S. J. & Scheller, W. G., (1993) *The Solar Electric House: Energy for the Environmentally-Responsive, Energy-Independent House,* USA: Chelsea Green Pub Co.

Stutzmann, M., (2009) *Handbook of Photovoltaics,* Wiley.

Thomas, R. & Fordham, M., (2001) *Photovoltaics,* UK: Spon Press.

Wenham, S, R.; Green, M. A.; Watt, M. E.; Corkish, R., (2006) *Applied Photovoltaics*, New South Wales: Earthscan Publications Ltd.

Periodicals
Home Power Magazine,
www.homepower.com

Renew
Alternative Technology Association
www.ata.org.au

Renewable Energy World
www.renewable-energy-world.com

Photon International
www.photon-magazine.com

Renew
NATTA (Network for Alternative Technology
& Technology Assessment)
http://eeru.open.ac.uk/natta/rol.html

Progress in Photovoltaics
www3.interscience.wiley.com/journal/5860/home

Organisations
BRE (Building Research Establishment)
www.bre.co.uk

Centre for Alternative Technology
www.cat.org.uk

Centre for Renewable Systems Technology
www.crestuk.org

Low Carbon Building Programme
www.lowcarbonbuildings.org.uk

Renewable Energy Association
www.r-p-a.org.uk

UK Photovoltaics Network
http://www.pvnet.org.uk/

PV system design resources
Retscreen – Free to download Excel spreadsheets,
handy for basic feasibility studies
www.retscreen.net

PV*Sol – Professional design application with detailed
system analysis www.valentin.de

PVSyst – Professional design package, very detailed and 3D shading modelling
www.pvsyst.com

Meteonorm – Global Meteorological Database supplied as PC application on CD-rom, includes simulation of shading, etc.
www.meteotest.ch

PV GIS – Free web-based resource, predicts system output for sites in Europe and Africa
sunbird.jrc.it/pvgis

Statutory regulations regarding solar PV

Electricity at Work Regulations – 1989

Building Regulations – 2003

Health & Safety at Work Act – 1974

Construction (Design & Management) Regulations – 1994

Electricity Safety, Quality and Continuity Regulations – 2002

The work at height regulations – 2005. SI2005 No.735, The stationary office

The work at height regulations: a brief guide
www.hse.gov.uk/pubns/indg401.pdf

Office of public sector information (formerly HMSO)
www.opsi.gov.uk

Non-statutory guidelines and regulations

Energy Networks Association (1989) Engineering Recommendations G59/1 – for grid connections above 16A per phase
www.energynetworks.org

Energy Networks Association (2003) Engineering Recommendation G83/1 – for grid-connected systems less 16A per phase
www.energynetworks.org

BS 7671:2008 Requirements for Electrical Installations (IEE Wiring Regulations 17th edition)

The institute of Engineering & Technology
(formerly the IEE)
www.theiet.org

Blackmore, P., (2004) 'Wind loads on roof-based
photovoltaic systems', *Digest* 489; BRE.

Blackmore, P; (2005) 'Mechanical installation of roof-
mounted photovoltaic systems', *Digest* 495; BRE.

Planning

Climate Change and Sustainable Energy Act 2006
http://www.opsi.gov.uk/acts/acts2006/ukpga_
20060019_en_1

Statutory Instrument 1995 No. 418 The Town and Country
Planning (General Permitted Development) Order 1995
www.opsi.gov.uk/si/si1995/Uksi_19950418_en_1.htm

(2004) Planning policy statement 22: Renewable Energy
http://planningportal.gov.uk/england/professionals/
en/1021020428382.html

(2007) Planning Policy Statement: Planning and Climate
Change, Supplement to Planning Policy Statement 1
www.communities.gov.uk/publications/
planningandbuilding/ppsclimatechange

The Merton (10% Renewable Energy) Rule
www.merton.gov.uk/udp/acrobat/udpfinal.pdf
http://themertonrule.org

Fuel security

Energywatch group (2007) *Crude Oil, The Supply Outlook*
http://www.energywatchgroup.org/Oil-
report.32+M5d637b1e38d.0.html

Climate change

(1992) *Kyoto Protocol to the United Nations Framework Convention on Climate Change*
http://unfccc.int/resource/docs/convkp/kpeng.html

Stern, N., (2007) *Stern Review on the Economics of Climate Change,* England: Cambridge University Press
www.hm-treasury.gov.uk/independent_reviews/stern_review_economics_climate_change/stern_review_report.cfm

Gore, A. & Guggenheim, D. (2007)
An Inconvenient Truth
US: Paramount Home Entertainment (DVD)
www.aninconvenienttruth.co.uk/

(2007) *Fourth Assessment report – Climate Change 2007: Synthesis report*
IPCC (Intergovernmental Panel on Climate Change)
www.ipcc.ch

Standards relating to solar PV systems

BS EN 14437 Determination of the uplift resistance of installed clay or concrete tiles for roofing,
Roof system test method

BS EN 61215:1995, IEC 61215:1993 Crystalline silicon terrestrial photovoltaic (PV) modules (monocrystalline and polycrystalline)

Design qualification and type approval
BS EN 61646:1997, IEC 61646:1996 Thin-film terrestrial photovoltaic (PV) modules

Design qualification and type approval
BS5534 Code of practise for slating and tiling (including shingles)

BS 6399-1 1996 Loading on buildings, Pt 1 Code of practice for dead and imposed loads

BS 6399-2 1997 Loading on buildings, Pt 2 Code of practice for wind loads

BS 6399-3 1998 Loading on buildings, Pt 3 Code of practice for imposed roof loads

British and some International / European Standards can be purchased from the BSI website www.bsonline.techindex.co.uk

Part P approvals bodies

BRE (Building Research Establishment)
Certification & ECA http://www.partp.co.uk

BSI (British Standards Institute)
http://www.bsi-global.com/en/ProductServices/Electrical/
Electrical-installers/

CORGI (Council for Registered Gas Installers)
http://www.trustcorgi.com/Consumers.htmx

NAPIT National Association of Professional Inspectors and
Testers
www.napit.org.uk/

NICEIC National Inspection Council for Electrical
Installation Contracting
http://niceic.com/

Distribution network operators

Eastern, London and South-East	EDF Energy	www.edfenergy.co.uk
South West	Western Power Distribution	www.westernpower.co.uk
South Wales	Western Power Distribution	www.westernpower.co.uk
North Wales	Scottish Power (MANWEB)	www.scottishpower.com
West Midlands	Central Networks plc	www.central-networks.co.uk
East Midlands	Central Networks plc	www.central-networks.co.uk
North and East Yorkshire	CE Electric UK (NEDL & YEDL)	www.ce-electricuk.com
North West	United Utilities	www.unitedutilities.com
Northern Ireland	NIE	www.nie.co.uk
South Scotland	Scottish Power (Central and Southern Scotland)	www.scottishpower.com
North Scotland and Central Southern England	Scottish and Southern	www.scottish-southern.co.uk

For full contact details, the Energy Networks Association publishes a list: www.energynetworks.org

Abbreviations and glossary

AC Alternating Current. An electricity supply which has an alternating voltage and hence current. All utility electricity distribution uses AC because voltage step up/down transformers cannot be used with DC.

Amorphous Refers to thin film silicon cells produced by depositing very thin strips of silicon onto a sheet of glass; uses much less silicon than crystalline technologies.

Amp The unit of electric current, symbol A.

AONB Area of Outstanding Natural Beauty.

Array One or more strings of solar modules installed on a roof, wall, pole, tracker or ground mounted frame.

Batten Thin horizontal strips of timber (usually 2"x1" (50mm x 25mm) onto which slates or tiles are nailed.

BERR Department of Business, Environment & Regulatory Reform (formerly the DTI).

Biomass Term used for electrical or heat energy generated from plant derived solid fuels.

BIPV Building Integrated Photovoltaics. A PV array mounted on a building as opposed to on the ground, or a vehicle, or boat.

Boost converter An electronic circuit which produces a DC output higher than its DC input (used because transformers do not work with DC).

BRE Building Research Establishment. Runs the government's micro-generation installer accreditation scheme; a registration body for Building Regulations Part P.

BSI British Standards Institute. A registration body for Building Regulations Part P.

BSRIA Building Services Research & Information Association.

Cell A single photovoltaic cell usually connected in a series string of cells to increase voltage to a useful level.

Charge controller A device which limits charging to prevent damage once batteries are fully charged; unit may combine load controller and other functions.

Conduit A round tube used to protect electric cables, usually PVC or galvanised steel.

CORGI Council for Registered Gas Installers. A registration body for Building Regulations Part P.

Current In an electric circuit, current is the rate of flow of electrons around a circuit.

DC direct current An electricity supply with a continuous voltage; electricity from batteries and solar cells is DC. *See also AC.*

DNO Distribution Network Operator. The distribution network is the electricity supply network within a given region which distributes electricity from the national grid to consumers.

DTI Department of Trade & Industry (now the BERR).

ECA Electrical Contractors Association. A registration body for Building Regulations Part P.

ELECSA A registration body for Building Regulations Part P.

ELV Extra Low Voltage (less than 120V ripple free DC or 55V AC).

Embedded generator *See SSEG.*

ENA Energy Networks Association (represents DNOs).

ETSU Energy Technology Support Unit, part of the BERR.

Feed-in tariff (FiT) When an electricity supply company is obliged to pay a minimum rate per kWh for renewable energy. Usually younger, higher cost technologies like PV attract higher feed-in tariffs than more mature technologies like onshore wind, for example, in Germany the tariff is over 40 euro cents per kWh for PV.

Grid-connection A source of renewable energy that is connected to the distribution network (hence this common term is slightly misleading since the system isn't directly connected to the national grid).

G59/1 Engineering recommendation for SSEGs above 16A per phase.

G83/1 Engineering recommendation for SSEGs below 16A per phase. Avoids the need for a DNO visit to inspect system protection, by using type approved inverters with factory default protection settings.

HMSO Her Majesty's Stationery Office (now OPSI).

Hybrid system A system supplied by more than one renewable energy technology, e.g. solar and wind or solar and hydro.

IEC International Electrotechnical Council. Issues standards for electrical engineering including PV products, tests and procedures.

IEE Institute of Electrical & Electronics Engineers (UK) (not to be confused with the US IEEE), renamed IET in 2007.

IEEE Institute of Electrical and Electronics Engineers (US).

IET Institute of Engineering & Technology (UK) (formerly the IEE).

Inverter Any device that converts DC electricity to AC. Also often steps up the voltage and performs other control, protection and monitoring functions according to application.

Irradiance The instantaneous solar power received on a given surface, measured in W/m^2 (watts per square metre).

Irradiation The cumulative solar energy received on a given surface over a specified period of time, measured in $Wh/m^2/year$ (watt hours per square metre per year), could also be per day or per month. The term radiation is sometimes used misleadingly since radiation refers to the energy emitted by a body (for example a lamp or the sun).

Islanding The situation occurring when an embedded generator continues to supply nearby loads during a power cut (loss of mains), which is potentially very dangerous as the grid will very likely be out of synch when it resumes, causing an explosion.

Isolator Usually refers to a disconnect switch that can be used to provide safe isolation of an electric circuit for maintenance rather than only functional purposes.

kWh Kilowatt hour A unit of energy, usually electrical, equivalent to a device that consumes or generates electrical power at a rate of 1kW for one hour. Also known as a unit of electricity and used on electricity meters and bills.

MWh Megawatt hour Equivalent to 1000kWh.

Laminate The term used for a PV module without an aluminium frame; often used for roof-integrated systems where the frame is supplied separately, usually from a different manufacturer.

Live parts Components that are electrically live and could cause electric shocks or fires in the event of a fault or accident.

Load controller A device that disconnects appliances and/or lighting when batteries are discharged ('flat') to prevent battery damage; a unit may combine a charge controller function with load control.

Lockable isolator An isolator that can be secured in the 'off' position with an electrician's own padlock whilst a circuit is being worked on.

Loss of mains Term used when the utility supply has failed due to a fault, aka 'power cut'.

Loss of mains protection A circuit built into G83/1 approved equipment that disconnects to prevent islanding.

MC Multi-contact A brand of PV cable and connectors.

Meter operator A company responsible for installing, owning, maintaining and reading the electricity meter for all buildings connected to the distribution network.

Micro-generation A small generator of electricity (and heat) located within a building, or other consumer of electricity. *See also SSEG.*

Module A module usually consists of a string of solar cells encapsulated in weatherproof glazing, with a junction box, sometimes factory fitted with cables and connectors.

Monocrystalline Silicon solar cells cut from a single silicon crystal.

MPPT Maximum Power Point Tracking. An automatic variation of the current drawn from a solar cell to optimise power output according to variations in temperature and irradiance.

NAPIT National Association of Professional Inspectors and Testers. A registration body for Building Regulations Part P.

National grid The high voltage electricity network designed originally to transmit electricity across the country (and continent) from large power stations to regional distribution networks.

NICEIC National Inspection Council for Electrical Installation contracting. A registration body for Building Regulations Part P.

NICd nickel cadmium batteries. A common type of rechargeable battery used in consumer goods and commercial applications, also known as NiCad.

Noggin A general term for a short length of timber fitted between two beams, often as a retro-fit, e.g. where a rafter has been cut to install a roof window.

Off-grid *See stand-alone.*

Ofgem Office of Gas and Electricity Markets. The government regulatory body for the industry.

OPSI Office of Public Sector Information, formerly HMSO.

PV Photovoltaic, any device which converts light directly into electricity.

Part P The part of the Building Regulations that applies to buildings.

PES Public Electricity Suppliers. The company that invoices for electricity sales; separated from DNOs during electricity privatisation in the 1980s.

Phase The phase conductor in a circuit is also known as the live conductor; a domestic or small commercial property will only have one phase conductor in the supply and in each circuit, so it is called single phase, whereas a large commercial or industrial building will have 3-phase as it is more efficient for high power systems.

Polycrystalline Solar cells cut from an ingot consisting of multiple silicon cells, giving it a crazed appearance.

PV/Photovoltaic The process by which a device converts photons of solar energy directly into electricity.

Rafter A diagonal timber beam, usually 3" x 2" (75mm x 50mm) or 4" x 2" (100mm x 50mm), onto which felt/ sarking and then battens are fixed; can be seen inside a loft space.

ROC Renewable Obligation Certificate.

SSEG Small-scale Embedded Generator. A generator of electricity that is connected to the regional distribution network.

SSSI Site of Special Scientific Interest.

Stand-alone In the context of renewable energy, a system that is not connected to the utility electricity network or national grid.

Tedlar™ Dupont trade name for the polyvinyl fluoride (PVF) used as a back-sheet in PV modules; has very good resistance to weathering and is very slow burning.

Tracker A mounting for a small array that pivots the array in one or two directions to follow both the sun on its east-west path and solar altitude.

Transformer A device that converts electrical energy into magnetic energy and then back into electric energy; used to step up or step down an AC voltage, or to provide separation between two AC circuits at the same voltage for safety reasons.

Trunking A rectangular channel used to protect electric cables, usually PVC or galvanised steel.

Tyco A brand of PV connectors, often supplied with modules.

Type approval When the final prototype of a device (in this case inverter) is laboratory tested to prove its compliance with a given national/international standard, avoiding the need for individual units to be tested on-site.

UNESCO United Nations Educational, Scientific and Cultural Organisation.

Volt The unit of voltage. This is the electromotive force that causes electric current to flow in an electric circuit, symbol V.

W/m² Watts per square metre. This is a unit of solar irradiance; values for the UK vary between 0 for absolute darkness to 1200 in very bright sunshine.

Wp Peak Watts. This is the unit for nominal DC output power of an array or module under standard test conditions ($1000W/m^2$, 25°C, etc.).

About the author

Brian Goss, MSc, was formerly the Electrical Engineer at the Centre for Alternative Technology (CAT). He studied Electrical Installation at Chesterfield College and Renewable Energy Systems Technology at the Centre for Renewable Energy Systems Technology (CREST), Loughborough University. See www.crestuk.org

Brian teaches on various solar PV courses at CAT, including the City & Guilds Certificate in Installing & Testing Domestic Photovoltaic Systems (C&G 2372); the MSc: Renewable Energy and the Built Environment, and also the short course 'Introduction to Solar Electric Systems'. See www.cat.org.uk/courses

Brian also works as Research Engineer for CREST on a PV systems optimisation project for the IKEA Group. The project will monitor large-scale solar PV arrays in real-world environments to analyse and improve the decision making process used to design solar PV systems.

In his spare time, Brian enjoys kayaking, mountaineering, ski-touring and cycling. Visit his website www.briangoss.co.uk

Acknowledgements

I would like to thank the following people:

Nick Mills (Dulas Ltd) for his valued suggestions.

Oliver Sylvester Bradley (Solar Century); Meryl Jones (Dulas Ltd); Fatmir Niqi (Solkraft Innovation); Peter & Andrea (SIT Europe): for their generous donations of system photographs.

Phil Horton; William Carey; Martin Ashby; Paulo Mellet: for their patience and assistance with case studies.

Allan, Annika, Caroline, Graham, Hele, Tobi, Lesley and Helen at CAT: for their patience and diligence through the various edits.

And of course Bettina.

The publishers would like to thank Savio Alphonso for allowing them to use his graphic on the front cover. www.savioalphonso.com